THE
DATAPRENEURS

THE DATAPRENEURS

THE PROMISE OF AI AND THE CREATORS BUILDING OUR FUTURE

BOB MUGLIA
WITH STEVE HAMM

PEAKPOINT
PRESS

Peakpoint Press books may be purchased in bulk at special discounts for sales promotion, corporate gifts, fund-raising, or educational purposes. Special editions can also be created to specifications. For details, contact the Special Sales Department, Skyhorse Publishing, 307 West 36th Street, 11th Floor, New York, NY 10018 or info@skyhorsepublishing.com.

Peakpoint Press® is a registered trademark of Skyhorse Publishing, Inc.®, a Delaware corporation.

Visit our website at www.skyhorsepublishing.com.

10 9 8 7 6 5 4 3 2 1

Library of Congress Cataloging-in-Publication Data is available on file.

Cover design by Brian Peterson

ISBN: 978–1-5107–7841-2
Ebook ISBN: 978–1-5107–7842-9

Printed in the United States of America

CONTENTS

GLOSSARY

Artificial general intelligence (AGI): AGI is a computing system with intelligence equivalent to an average person.[1]

Complex data (Unstructured data): "Complex data" describes content like text, video, still images, and audio, which remain challenging to manage using traditional databases.

Data applications (intelligent applications): Unlike most conventional applications that respond to people, a data application takes independent action in response to changes in data.

Data cloud: Most of the digitized data in the world is mass-migrating to the public cloud, where service operators run remote data centers. Data is managed, shared, analyzed, and safeguarded more easily there.

Data economy: This twenty-first-century phenomenon dictates how data and data analytics (including AI) increasingly power economic growth and vitality.

---

1 Wikipedia, 2004. "Artificial General Intelligence." Last modified March 2023. https://en.wikipedia.org/wiki/Artificial_general_intelligence.

Foundation models: Large-scale machine-learning models created and trained using the entire universe of information found on the internet.[2]

Generative AI: Applications built on foundation models that generate output and provide well-defined services. ChatGPT, DALL-E, and Stable Diffusion are frequently called generative models or generative AI.

Information at your fingertips (IAYF): An aspiration identified by Bill Gates and other leaders at Microsoft in the early 1990s to make information more readily available on personal computers.

Knowledge graph: A technology that manages the digital representations of things, events, situations, and concepts. It is typically implemented using a database that can model concepts and the relationships between them. Knowledge graphs are often visualized using a graph structure.

Modern data stack: A collection of technologies developed by many companies in the past ten years to support data management and data analytics in the cloud.

Relational knowledge graph: A knowledge graph that is based on relational mathematics. Relational knowledge graphs enable the granular modeling of both concepts and relationships, as well as the definition of rules and constraints. Relational knowledge graphs are executable models that bring code and logic together into a single database.

Superintelligence: An evolution of artificial general intelligence that is smarter than all humanity combined.

Technological singularity: A future time when superintelligence enables hundreds of years of progress in one year.

---

2   Wikipedia, 2022. "Foundational Models." Last modified March 2023. https: //en.wikipedia.org/wiki/Foundation_models.

PREFACE

Even as a youngster, though, I could not bring myself to believe that if knowledge presented danger, the solution was ignorance. To me, it always seemed that the solution had to be wisdom. You did not refuse to look at danger, rather you learned how to handle it safely.
—Isaac Asimov, "The Caves of Steel," 1953

———

The idea of writing a book started a few years ago while I was still CEO of Snowflake. It was and is one of the most successful startups in cloud computing—where servers, networks, storage, and applications reside in vast data centers accessible through the internet. Snowflake's chief marketing officer Denise Persson had the brilliant idea to publish a book before our initial public offering (IPO) about the company, our founders, and the magic of managing data in the cloud. Since I was the CEO, I was the obvious choice as author.

Because of the time commitment, I was a bit reluctant at first. However, Denise convinced me, and I became excited about the idea. I am passionate about helping organizations manage and analyze their data—growing businesses and seeding the data economy. I wanted to tell that story. Snowflake hired Steve Hamm, a veteran tech reporter and book author, to help. Steve and I readied to embark on the adventure when, surprisingly to me, the Snowflake board replaced me in April 2019 with another executive who had deep experience in taking companies public.

That, it seemed, was that. As you will read in these pages, I went on to reinvent myself as an investor and adviser to startups in the data management space. Meanwhile, Steve coauthored the Snowflake book, *The Rise of the Data Cloud*, with the company's new CEO, Frank Slootman. After leaving Snowflake, I had the opportunity to talk with many experts and, in some cases, invest in companies that provide vital technologies for the modern data stack. I hoped to write about companies I remain involved with and what I learned along the way. Through intense conversations with executives and scientists, I realized I had a bird's-eye view of data's future and its impact on businesses and the world.

Initially, I wanted to contribute short and pithy articles to tech publications. I reached out to Steve, who had completed the Snowflake book by then. Initially, Zoom meetings with Steve consisted mainly of me giving him brain dumps. I wanted to share all my thoughts and learnings about the present and future of data. To manage the deluge, Steve started writing a narrative. Over time, as the story grew, we saw the potential to produce a memoir-style book about my life with data.

As we talked and I looked back on my career, I realized that although I spent over 20 years at Microsoft, I worked with entrepreneurial people from the beginning. I had the good fortune to work with many gifted technologists and data visionaries who drove my thinking and career. I told Steve I wanted the book to be about these amazing "data entrepreneurs," and he quickly came up with the title *The Datapreneurs*.

Writing this book prompted me to think deeply about my life and the future of technology and humankind. While I am not religious in a frequent-attendee-of-services sense, I embrace the values taught to me in my childhood and strive to live my life accordingly. I believe what matters is the impact each of us has on the world around us.

People serve humanity in many ways, but my top motivator is inventing and building things. I am passionate about learning, creating, teaching, and doing my best to contribute to humanity's knowledge. I love to work with others to build new technologies that positively impact our lives. How we do this matters and should be rooted in our values. To me, it means acting with integrity and respecting and honoring those around us. Once we achieve success, we should give back to society.

Accordingly, this book became part memoir and part history of the people and technologies that made the data analytics era possible. It is one way for me to give back by teaching others about data and technology's astounding capabilities. In this book, I want to convey some of my most crucial technology observations and describe how ethics and values play a critical role. I also seek to share my stories in a fun and easy way to avoid creating a dry technology primer.

But as Steve and I developed this story, the world changed rapidly around us, and technological advances came faster than I had thought possible. Recent advances in computer science catapult us into an era of ever-smarter machines that can hyper-accelerate scientific and economic progress. The implications are mind-blowing. And so *The Datapreneurs* looks also toward the future of this technology.

I want to acknowledge a critical issue of our times up front. I have written quite a bit about artificial intelligence (AI) and how specially designed computers can mimic humans. AI has tremendous potential, both for good and ill. It is a hugely hot topic: Practically every day, it seems, another article is published about advances in generative AI, chatbots, and the promise of artificial general intelligence[3] (AGI)— along with warnings about the potential harm that could come to humans.

---

3 Wikipedia, 2004. "Artificial General Intelligence." Last modified March 2023. https://en.wikipedia.org/wiki/Artificial_general_intelligence.

Technologies underlying AI, including machine learning, have the potential to assist us toward healthier, more fulfilling, and more sustainable lives. At the same time, AI could serve in antisocial and even antihuman ways. Therefore, AI comes with profound ethical issues to consider. I'm not an ethics expert, but I want to convey the necessity of embedding ethics and values into today's and tomorrow's computer systems. I also want to share some prescriptive guidance.

The importance of these issues became clear to me in 2022. After seeing the GitHub Copilot service—which helps developers write code—I had my first aha moment. Other aha moments resulted from introducing new services, including DALL-E 2 and ChatGPT. These innovations drove me to read and watch everything I could find on foundation models,[4] which are machine learning models at an enormous scale. Foundation models tap the universe of information on the internet for their creation and training. As I talk to colleagues across the industry, I find they are also awed by the implications of this new technology.

Fascinatingly, the machine-learning techniques used to develop and improve these models demonstrate emergent capabilities—in other words, they demonstrate abilities their designers did not anticipate. At the same time, some of their responses to queries produce wrong or even ridiculous answers we call "AI hallucinations."[5] Over time, we will learn how to correct these mistakes.

We are on the verge of breakthroughs in machine intelligence and will likely see AGI achieved during my lifetime. Today, technology evolves so rapidly that we will probably see the introduction of many new products by the time you read this book.

I believe that tech business leaders like me, ethicists, and policy leaders must collaborate to seek answers to the short- and long-term questions

---

4   Wikipedia, 2022. "Foundational Models." Last modified March 2023. https://en.wikipedia.org/wiki/Foundation_models.
5   Wikipedia, 2022. "AI hallucinations." Last modified March 2023. https://en.wikipedia.org/wiki/Hallucination_(artificial_intelligence).

raised by the emergence of AGI. It may take longer than I predict, but AGI is undoubtedly coming. And we had better get going because advances are coming so fast that society risks getting caught flat-footed.

I have always believed that people and technology can ultimately solve any problem. While writing this book, that belief grew stronger, and I realized that I am a humanist.[6] We are all in this together, and I believe people are the solution to any problem. I am also a techno-optimist. I recognize the incredible challenge of governing machine intelligence. We will make mistakes along the way, but I believe we will eventually get it right. My hope is that *The Datapreneurs* is one small contribution toward that end.

A handful of other disclaimers:

- This book is about the people and companies I know. Many other individuals across the industry and academia have made significant contributions in this field and are only briefly mentioned in the book or not acknowledged. In particular, every datapreneur has support from a team that delivers on the vision and product. All of these people play critical roles, and their contributions are invaluable.
- I am extremely lucky in my life and career. In so many ways, I have been fortunate to be in the right place at the right time. I make no grand claims about my contributions to advances within the domain.
- I am not objective. I own stock in a number of the companies I talk about, so I have skin in the game.
- Also, I bring my values and preferences to this project. I see data's past, present, and future through Bob-colored glasses.

So, if you are willing, please read on.

---

6 American Humanist Association, 2003. "Humanism and Its Aspirations: Humanist Manifesto III, a Successor to the Humanist Manifesto of 1933." https://american humanist.org/what-is-humanism/manifesto3/.

INTRODUCTION

One of my favorite mementos kept in my office is a sealed box of the first significant product I was involved with at Microsoft, where I worked for twenty-three years. The box contained the Ashton-Tate/ Microsoft SQL Server, the first database management system that Microsoft sold for the newly emerging client-server software market.

The box and the text on it show how far we have come in the computer industry since 1989, when the product was released. It contained seventy-two floppy disks and upward of ten pounds of user manuals. What a headache! Now, think about today, in the cloud computing era. Computer operations can run in massive cloud data centers owned and operated by somebody else. Thanks to the cloud, somebody else handles all that loading and software management rather than the end user. As a result, businesses large and small have ready access to immense amounts and types of data, and they can use it to provide superior products and services and to fuel their revenue growth and profits.

That profound shift to cloud storage has enabled the emergence of the data economy, promising a future powered by artificial intelligence. Today, many businesses understand that data is one of their critical assets, perhaps second only to their employees.

Today's global economy—including the cumulative output of all of humanity—is powered by information. However, it also relies on natural resources like coal, which propelled the industrial revolution, and oil, which fueled the explosion of productivity and wealth in the twentieth century. It seems likely that data and artificial intelligence, together with alternative types of energy (wind, solar, nuclear, perhaps even fusion), will team up to power the twenty-first century. The combination of human and machine intelligence is on the verge of utterly transforming the world.

We are in a remarkable new era of computing and business where data rules. Thanks to the cloud, new data management software, and machine learning technologies, we can access an immense amount of data and information. This data can be easily accessed, integrated, managed, shared, bought, sold, and investigated. This reality fulfills Bill Gates's idea shared at COMDEX, a huge computer industry trade show, during his keynote speech in 1990. He predicted that personal computers (PCs) and software would put information at everybody's fingertips, and he was right.

In the coming years, it seems likely that organizations of all sizes—even individuals—will be able to tap tremendous cloud-based computing, storage, and, ultimately, knowledge and intelligence at a low cost. We will harvest data from devices, applications, video cameras, and cloud services. Data will enable new generations of artificial intelligence services to act as assistants. These AI assistants will help us understand the world around us better, improve the economy, increase human well-being, and predict the future.

Humanity's journey to take full advantage of this data has just begun. There is so much more we need to do technologically to make these systems work simply and efficiently and to make them available to large numbers of people—not just data scientists and high-end business analysts. At the same time, automation will take over many routine, human-driven processes. These new capabilities will help confront today's challenges, including climate change, overpopulation, the spread of infectious diseases, and profound social stresses.

Many factors enable the emergence of the data economy and artificial intelligence. Cloud computing is a major one. The explosion of data from a wide variety of sources is another. But I want to focus next on the virtues of the data-handling ecosystem that has taken shape over the past ten years in tandem with the rise of cloud computing—with tremendous progress coming in the past three or four years. The World Wide Web took off in the 1990s as a powerful vehicle for communications, media, search, and buying and selling merchandise. Still, it was not until after Amazon introduced Amazon Web Services (AWS)—with Microsoft and Google ultimately following—that the cloud became an essential platform for data management and analytics. The cloud theoretically made all global data of every type available to anybody who wanted it.

Before broad cloud adoption, a tsunami of digital data quickly overwhelmed the computing systems of the day. On-premises databases and corporate data centers became inundated. Specialized computing systems that could handle large quantities of data were too expensive. Plus, data was difficult to access since portions spread across scattered server "islands." Later, products based on Apache Hadoop helped people manage and analyze data using large clusters of computers. While Hadoop had a cute elephant mascot and logo, it was not a very nice elephant. Hadoop was too complex for all but the most technology-savvy organizations.

In those days, most data analytics were performed on personal computers using Excel spreadsheets, with different people having ownership of various chunks of data. There was no easy way to work with all the appropriate business data. Instead, it was split apart into the hands of many different people. The spreadsheet is a tremendously useful piece of software, but it is not up to the task of processing vast amounts of diverse data.

The result was data chaos.

Fortunately, a new generation of technology helps organizations manage and analyze our wealth of data: the modern data stack. The widely used term means different things to different people, which can be confusing. That is why I spent some time developing a crisp definition:

The modern data stack is data analytics delivered as a software service. It leverages the public cloud for scale and low cost and facilitates data modeling with SQL databases. For the first time, business customers of all sizes can gain insights using modern analytics delivered as a software service at an affordable cost.

The modern data stack is an ecosystem of technologies provided by many companies and open-source software projects. Nobody owns or controls all of it, so innovation abounds.

This ecosystem is powerful in its present state. It enables people to organize, access, and integrate data more conveniently and affordably. Unsurprisingly, much of the world's data is migrating to the cloud and the modern data stack, making it easier to access and manage. The modern data stack also makes it incredibly easy for organizations to break down their old data silos and share data internally and with their business partners and customers.

But the modern data stack is not yet mature. Some things need improving, and other things are simply not possible yet. We are in the early days of machine learning and foundation models. Over the next few years, advancements will change every application and hugely impact businesses and consumers.

You can think of this future as a model-driven world. The models will essentially be digital twins of real-world businesses. These models will be described and continuously improved by consumers and business leaders, encoded by software programmers, queried by data scientists, and constantly optimized using artificial intelligence. The feedback loops between the models and the organizations will drive continuous improvement, creating more successful and sustainable businesses, industries, and social systems.

Writing this book is one way I can help fulfill this vision. By describing my career journey in the tech industry and exploring the significant computer science advances that reshape our world, I hope to help you understand the past, present, and future of data as I do. I aim to share my experiences and insights about designing and building great organizations and world-changing products over the course of more than forty years in the tech industry.

I experienced some hits and misses in products and technologies. Late in the 1970s, while studying at the University of Michigan, I worked for a small relational database consulting company, Condor Computer. Since then, I have been fascinated by how to gather, store, and use data. That fascination intensified in my years as a tech industry executive, first at Microsoft, then at Juniper Networks, and finally as CEO of Snowflake.

I helped produce some fantastic products at Microsoft, including SQL Server, Visual Studio, Windows NT, and Systems Center. Then, at Snowflake, I helped the company grow over five years from zero revenue and thirty-three staff to annual revenues of $200 million and 1,300 employees. But I have also made my share of mistakes. I was involved in the development of products that were ahead of their time, such as Netdocs, an ambitious, early attempt to convert Microsoft Office to run on the web. Others were off target, such as the version of Windows code-named Cairo, which tried (and failed) to build Bill Gates's vision of "information at your fingertips" into the operating system.

At Microsoft, I worked side by side with Bill Gates and a group of technologists on a mission to democratize computing power and information access. My time at Juniper Networks helped me appreciate connectivity's significance more fully. The first two founders of Snowflake, Benoit Dageville, and Thierry Cruanes, taught me how brilliant, dedicated founders can create a thriving company. My time there solidified my belief that we should never separate technology from values and ethics. The companies I worked for—and my interactions with many entrepreneurs—helped me recognize the importance of values in a business organization. Creating and selling products, and building your engineering and sales teams, are the "what" of doing business. Values are the "how."

Since I left Snowflake, my passion has taken a new path. I am investing in and helping startups that fill the gaps in today's data ecosystem and invent future data ecosystems. In that role, I see companies build on great ideas that have long driven the evolution of computer science, starting with the many diverse types of data and the ways of processing it. I have advisory roles, investments, and, in some cases, board memberships with several companies advancing technology and driving the future of the modern data stack. They include Microsoft, Snowflake, RelationalAI, JuliaHub, Fivetran, Fauna, Pinecone, and Docugami. My vision is a roll-up of their visions. They are building the future of data management, and I am helping as an adviser and, in some cases, an investor or board member. In keeping with the Darwinian nature of early-stage entrepreneurship, things do not always go as planned, so some of the companies I discuss in the upcoming chapters may not succeed. Regardless, I continue to learn from these entrepreneurs. Being along for the ride is gratifying, and, in my small way, I can help propel things forward.

As you will see, a major part of this project is using my experience to trace the arc of data innovation and technology adoption that brought us today's data economy and modern data stack. This arc represents the continuous progress of innovation that is now hurtling us into the era of artificial general intelligence.

Visual representations are often the best way to show information. In early 2023, I sat down and traced the arc of data innovation shown on page 7 to help readers appreciate the progress from 1950 to today and extending into the future.

Each label on this arc represents one of the most critical data innovations that led us to today. This book highlights these innovations and some of the datapreneurs that made them possible.

Progress is now advancing at what appears to be an exponential rate. We seem close to building AI assistants that achieve artificial general intelligence (AGI), producing machines as intelligent as people. But many challenges still exist, so I expect progress to happen in stages.

We will introduce AI into applications and interact with them in English and other native languages. The internet search process will transform first since it plays a central role across different tasks and is

The Arc of Data Innovation

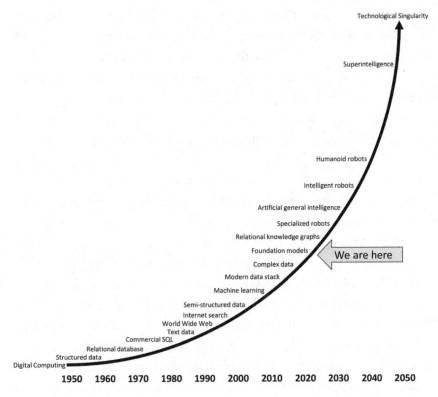

the most valuable application on Earth. Search and other AI assistants will become more intelligent over time and will do more things for us. As people engage with these AI assistants, the AI will learn from us and gain intelligence. Within ten years, I predict that some of these AI assistants will pass the threshold of median human intelligence and will be considered an AGI.

It won't stop there. These AGI assistants will continue to learn, and the pace of that learning will increase. Progressive generations of AGI will likely develop into machines with artificial superintelligence, making them more intelligent than all humans combined. Further on, futurists predict mass deployment of AGIs throughout society (which some people call the technological singularity) when, ostensibly, we can harness universal computer intelligence to rapidly

accelerate innovation, allowing us to address many of today's seemingly intractable challenges.

The combination of human and machine intelligence could start
an era of progress and prosperity, ultimately resulting in a world of
plenty where everyone on Earth can have what they need and want.
This vision is compelling. Let me explain. Our economy has three
essential elements: intelligence, energy, and labor. AGIs and artificial
superintelligence will add intelligence to applications and devices
of all types, ultimately lowering the cost of this intelligence to near
zero. Technology advances in wind, solar, nuclear, and, perhaps, fusion
could radically reduce the cost of electricity. Mass-produced smart
robots will, over time, crush the costs of certain types of labor. If they
follow the course I hope for, these forces could reshape society and
individual existence in ways we cannot yet imagine.

Right now, I imagine that many of you are hearing alarm bells
going off in your heads. Bob seems to be smitten with magical thinking. Point taken. I may have fallen too hard for these optimistic visions
of a better world just around the corner. Still, I am a realist. I recognize that technological advances often take longer—and with greater
difficulty—than expected. They also typically bring unexpected challenges along with their benefits. The time frame for future advances
is uncertain.

Perhaps more importantly, there are many alternatives to the rosy
picture I paint. Dystopian science fiction, literature, and various articles explore many possibilities.

To help with some of these questions, I look to my favorite science
fiction author, Isaac Asimov. Starting in the 1930s and continuing to
the 1990s, he authored over 460 books and short stories, including
his autobiography, *I, Asimov: A Memoir*. He was an optimist and a true
visionary. His moral insight and ability to foretell the future make him
nothing less than a prophet.[1]

---

1   Merriam-Webster.com Dictionary, s.v. "prophet," accessed March 27, 2023, https:
    //www.merriam-webster.com/dictionary/prophet.

Asimov understood the capabilities of early digital computers and imagined distant futures where intelligent machines played ever-more-important roles in society and the lives of individuals. He viewed robots as essential tools for human expansion into space, our solar system, and, eventually, thousands of worlds throughout the galaxy. Asimov predicted that some people would fear these robots and reflected that sentiment in many stories.

While Asimov did not get everything right (Who could?), he was amazingly prescient in some ways. Throughout his massive body of work, Asimov viewed humanoid robots as machines that are programmed to serve people, and he examined the ethical issues this raises. Starting very early on, Asimov recognized that society would need guidelines to regulate intelligent machines' design and control their behavior. He even mapped out rules called Asimov's Laws of Robotics. I will explore these in detail in Chapter 12.

I think of Asimov's Laws as something akin to the Jewish and Christian Ten Commandments. They are solid rules to live by, but they are insufficient to deal with the challenges of our era, when artificial general intelligence (AGI), the ability of machines to match or even surpass human intelligence, seems to be within our grasp. Indeed, we have reached a point in human history where it is necessary to create a new kind of social contract that governs our relationships with intelligent machines and helps society and individuals deal with the rapid changes and disruptions that will likely come.

That is part of the remit of *The Datapreneurs*. I will take you on a guided tour of the history of data management and analytics as I have experienced it and how I see it. To make your journey interesting and fun, I tell the story primarily by describing the datapreneurs in the computer industry who made some of the most important contributions to the evolution of data processing. These datapreneurs include people I listed in the preceding paragraphs, some less-familiar individuals who made critical contributions in the past, and others whose brilliance is yet to be recognized. And I conclude the tour with our AGI future and the new innovations that enable it. I'll propose how we might encode ethics models within our future intelligent machines to ensure they contribute to the betterment of humanity and fulfill the new social contract we need.

PART ONE

THE RELATIONAL REVOLUTION

The relational database transformed the computer industry, making it easier to manage and retrieve data and write run-the-business applications. In combination with PCs, relational databases put information at people's fingertips. It also paved Microsoft's way into the world of enterprise computing.

CHAPTER I

MICROSOFT MOVES ON THE ENTERPRISE

In the early days of the PC industry, new products drove major shifts in the industry narrative. Ashton-Tate/Microsoft SQL Server was one of them. On January 13, 1988, in the ballroom of the Marriott Marquis New York Hotel, Bill Gates took the stage with Ed Esber, CEO of Ashton-Tate, and Bob Epstein, one of the founders of Sybase. They announced a partnership to adapt Sybase's relational database to run on the OS/2 operating system jointly developed by Microsoft and IBM. They planned to sell it through all three companies' distribution channels.

Although the OS/2 operating system and Microsoft's relationship with Ashton-Tate would ultimately fail, the product introduction created a turning point in computer history. It marked the first significant incursion of the PC industry into serious business computing. It kicked off a multiyear effort culminating in Microsoft's version of SQL Server becoming the best-selling database server product of all time.

At the time, Microsoft rode high on the popularity of its MS-DOS PC operating system and Excel spreadsheet. Windows 1.0, introduced in 1985, caused barely a ripple, and Microsoft decided to join forces with IBM to codevelop OS/2 for the business market. (From its earliest days, Microsoft focused on working with computer-maker partners like IBM and Compaq.)

On that day, we hoped to formalize a partnership with Ashton-Tate, then the leading seller of databases for desktop PCs, and Sybase, a fast-growing provider of databases for client-server computing. The connection would give Microsoft a foothold in corporate computing, then dominated by IBM and Digital Equipment Corp.

Bill Gates and Ed Esber did all the talking onstage that day in the Marriott ballroom. Bob Epstein's role was to stand with them and "smile," as he remembers it. Bob had a good reason to be smiling that day. Sybase and the product he created were the most crucial technology players on the stage.

Bob was a pioneer of the relational database. In the 1970s, as a PhD candidate at the University of California, Berkeley, Bob managed the INGRES project, one of the first efforts to create a relational database in an academic setting. He later helped launch two startups built upon the relational concept, the second being Sybase. The Microsoft/Ashton-Tate/Sybase connection was a big deal, he says, because it democratized the relational database. "It brought the cost of computing way down and, all of a sudden, smaller companies could run the business on computers," Bob recalled recently.

Bob's adventure with Bill Gates began about eighteen months earlier when he received a call from Rob Glaser, a Microsoft executive. Bill wanted to talk with Bob. Though it was a small company at the time, Sybase challenged some of the giants in the database market with a server product based on SQL. Its architecture made efficient use of computing and network resources.

Microsoft had recently gone public and proved the computer software industry's viability. Bill visited Sybase in Emeryville, California, for the first meeting. The depth of Bill's understanding of databases surprised and impressed Bob. After several subsequent meetings, the two companies decided to work together. Bob recalls thinking: "Suddenly, we're in the big leagues."

My first encounter with Microsoft was also a call from Rob Glaser. In late 1987, I worked for ROLM, a telecommunications equipment pioneer based in Silicon Valley. There I met my wife, Laura Ellen, a Stanford MBA graduate, and we planned our life together. Laura Ellen envisioned the PC industry as the future of computing and applied for a job at Microsoft. She ultimately worked in systems marketing, and Rob learned about me through her. Rob called and interviewed me for a new technical role at the company—program manager for SQL Server. I became the second employee on the SQL Server team.

I remember my first impressions of Microsoft. I joined the company in early January 1988, just a few days before the Ashton-Tate/ Microsoft SQL Server announcement. At the time, Microsoft employees resided in a cluster of six small buildings in suburban Redmond, Washington. I worked in Building 2 near the modest-sized pond that employees had named "Lake Bill." I had a tiny interior office—befitting my lowly status—but I barely noticed. Learning about SQL Server and preparing a demo for announcement day kept me extremely busy.

I worked previously with several databases, including an early relational database called Condor, which ran on a Cromemco microcomputer that predated the IBM PC. This computer used a Z-80 microprocessor and a mind-bogglingly small sixteen kilobytes of memory. However, most applications I worked with before joining Microsoft depended on hierarchical and network databases that preceded the relational era.

In the 1940s, the first digital computers that could store both data and a program containing executable instructions emerged. The 1950s saw the introduction of the first commercial computers. These early computers relied on paper or magnetic tape for long-term storage. Magnetic disk storage emerged in the mid-1960s and enabled dynamic access, which allowed data retrieval and storage on demand and made databases possible. In 1968 IBM introduced the Information

Management System (IMS[1]), one of the first successful early databases. IMS ran on IBM mainframe computers and supported the incredibly complex parts list for NASA's Apollo moon rocket project. IMS used a simple tree hierarchy for organizing information.

Hierarchical structures are a widespread and advantageous way to organize data. Nature is often hierarchical. We witness it whenever we see a tree's branches and leaves or when we look at a river formed by smaller tributaries. Because a hierarchal structure is natural to us, we commonly see data stored that way in file browsers and email applications.

While hierarchies are instrumental for data organization, something else must maintain the relationships or links between things. Early business applications used "network" databases to overcome this limitation. Network databases stored connections between records, allowing for much more complex data structures.

Both hierarchical and network databases were prevalent during the 1970s and the first half of the 1980s, but they had some serious problems. These systems had rigid architectures, making it complicated to create reports. They used hard-coded links, which were awkward to program and difficult to update.

Ted Codd, an Englishman who worked at IBM Research in San Jose, had a better idea. In 1970, in an essay, "A Relational Model of Data for Shared Data Banks,"[2] he laid the groundwork for today's relational databases by building on a foundation of relational mathematics. Relational databases organize data into tables with keys to identify each row. A relational database can define the relationships between different tables at any time. This capability eliminated the hard-coded links between data, providing considerably more flexibility in application design.

1   Wikipedia, 2003. "IBM IMS Database." Last modified February 2023. https://en.wikipedia.org/wiki/IBM_Information_Management_System.
2   E. F. Codd, "A Relational Model of Data for Shared Data Banks." *Communications of the ACM*, 1970. https://www.seas.upenn.edu/~zives/03f/cis550/codd.pdf.

Because relational database designs use the principles of relational mathematics, it is possible to specify a command to tell the database what to do without providing detailed instructions. This approach differed from hierarchical and network databases, which required application developers to provide specific instructions to retrieve and store data.

Initially, IBM resisted the relational approach, but common sense prevailed, and in 1977 IBM introduced System R, the world's first commercial relational database. System R introduced a new query language: structured query language, or SQL for short. SQL provided a powerful and easily understood way to specify a task for the database and enabled very complex queries using a single command.

Unlike the rigid, fragile approach of previous hierarchical or network databases, SQL allowed developers to work flexibly with data and change the structure over time. With SQL and other relational query languages, data could group into tables and relate to other tables using simple commands.

SQL also made it very easy for developers to group a set of updates into a single command called a transaction. Transactions are essential for any database, and SQL makes it much easier for developers to implement them properly. The classic example of a transaction is a debit-credit for a fund transfer between financial accounts. Transactions ensure that both the debit and credit happen simultaneously, or, if not, they both fail consistently.

IBM pioneered SQL with System R and later introduced its highly successful DB/2 database (now called Db2). Oracle was an early adopter of SQL, and founder and longtime CEO Larry Ellison built an industry giant based on the success of the Oracle database.

Bob Epstein recalls that in the 1980s, business computing was the domain of high priests of technology, a relatively small clique of experts who knew their way around hardware systems and databases. Only the most prominent companies could afford to hire them. Both IBM and Oracle focused on selling products to large enterprises. Bob took a different approach.

Bob's path to building a disruptive database started when he read Ted Codd's essay as a graduate student at Berkeley from 1977 to 1980. Bob was completing his master's degree when Prof. Michael Stonebraker, his thesis adviser, invited him to become the third manager of the INGRES project, essentially paying him to obtain his PhD. Bob said, "yes." He went on to write a lot of INGRES code and supervise a small team of graduate student coders.

In the INGRES project, he saw a way to democratize the power of computing—so anybody could do it. He decided to test the database in the real world and used it to create an application for running his wife's restaurant, All's Fare, in nearby Point Richmond. He did that over the weekend. "Computing was expensive and inaccessible," Bob recalls. "I thought all of that was about to change. We could do it at a fraction of the price and change an industry."

And change an industry they did. SQL databases saw wide adoption because they made it much easier for developers to build applications for businesses of all sizes.

In the 1980s, several new relational databases emerged. The startup spirit swept through the Bay Area, and work on INGRES at Berkeley spawned several companies. After Bob got his PhD, he played a pivotal role in two database startups, the second of which was Sybase. In 1984, he convinced Mark Hoffman, Tom Haggin, and Jane Doughty to join him in starting the company. Tom and Jane were the technical team. As CEO, Mark brought valuable business and manufacturing skills. Bob filled in many gaps as CTO and a jack of all trades. He designed software, set up customer training classes, raised capital, and helped with marketing and sales.

Sybase stood out from other database startups because it focused on designing a system to run on a network of computers rather than a single computer. Computer networks emerged, ushering in a new era of connected, lower-cost microprocessor-based computers. These systems employed a new graphical interface, making it much easier for people than the old command-line interfaces of mainframe and minicomputer systems. In addition to Microsoft's work with PCs, companies like Sun and Apollo built powerful computers that would play an essential role in business systems during the 1990s. New SQL databases and graphical user interfaces built on the Unix or Windows operating systems defined the "client-server" network orientation.

Sybase's relationship with Microsoft provided the ultimate stamp of credibility. Initially, Microsoft did not secure the rights to use Sybase's source code. That protected Sybase's intellectual property from a partner that would progressively become more of a competitor but left Sybase with the burden of handling all the customer support calls for service. So, Bob and his colleagues agreed to modify the contract. Ultimately, that shift opened the door for Microsoft to develop its database software based on the Sybase source code. Over time that product became a monster—Microsoft SQL Server. I will tell you more about that in Chapter 3.

Sybase had a nice run. Its product was very effective at handling business transactions and widely adopted within the financial services industry. The company had a successful public offering in 1991. However, a few years later, it stumbled as the market shifted. Customers started purchasing applications rather than buying a database and building them. After its purchase by the German software giant SAP, Sybase's software legacy lived on. Bob says Sybase's problems were "self-inflicted" and not caused by Microsoft. He has fond recollections of days when relational databases ran on client-server networks. "We were there for a major market transition," he says. "When you come in with the right product at the right time, you have a shot."

In many ways, the 1980s were the golden years of discovery for SQL databases. Companies like Sybase and Oracle demonstrated the superiority of SQL and the relational approach. The work that Michael Stonebraker, Bob, and others did at Berkeley continues to resonate, and modern SQL systems derive from that early work. PostgreSQL, wildly popular on today's cloud platforms, can trace its lineage directly back to Michael Stonebraker and those early days in Berkeley.

Bob Epstein did not have a speaking role at the product announcement of Ashton-Tate/Microsoft SQL Server in 1988, but his baby, SQL Server, was center stage. Bill Gates and Ashton-Tate's Ed Esper had all the business muscle in the three-way partnership. But, in the end, it was superior technology developed by Bob and a generation of database engineers that launched the revolution that powers modern business.

I was a peripheral player in my early days as a SQL Server program manager at Microsoft, but my experiences at that time put me on a path that has sustained me throughout my career. Early on, I recognized the importance of data for business and society. I was not a great software programmer, but I worked with incredible people. Through them, I developed a deep understanding of the technologies underlying modern data management and analytics.

I was fortunate to have a ringside view of the hugely consequential technology revolution that relational databases and SQL would bring to the modern business world.

CHAPTER 2

INFORMATION AT YOUR FINGERTIPS

Bill Gates has been an icon of global business and technology for so long that it is hard to remember that Microsoft was a relatively small company in the 1980s. Bill's influence existed within the relatively small orbit of the personal computer industry. These were the early days of the PC revolution. During the 1980s, the industry visionaries included the likes of Apple's Steve Jobs, Lotus' Mitch Kapor, and Intel's Andy Grove. Bill was very much in the mix, too, but it was not until 1990 that he took on the role of chief visionary for techdom. That elevation in status came when he introduced the goal of computers providing "information at your fingertips" at the COMDEX computer trade show. At Microsoft, we called it IAYF for short.

The IAYF concept has continued driving the tech industry ever since then. The vision remained simple. When individuals can access all kinds of information at the press of a button, they can live richer lives, be more successful in personal and professional pursuits, share information, and collaborate more successfully. That foresight helped transform the PC from a hard-to-use device with limited utility into humanity's indispensable electronic appliance until the smartphone took that role. It also helped drive the efforts to transform remote islands of data into consolidated, actionable information.

Let me take you back to Nov. 12, 1990, when Bill publicly used the "information at your fingertips" label and laid out his vision. COMDEX was the giant annual computer industry trade show that took place each year in Las Vegas. During the 1980s and 1990s, it was the must-do event for every company in the tech industry, and the keynotes were avidly anticipated and closely scrutinized. Bill's previous COMDEX keynote had come in 1983. Back then, COMDEX still had amateurish touches. Bill's dad ran his slideshow. By 1990, Bill's keynote was a much more polished affair. Microsoft communications strategist Jonathan Lazarus hired a filmmaker to produce a series of vignettes showing actors using PCs in various scenarios. They also included laugh-line references to the then-popular *Twin Peaks* television series.

Bill surprisingly started his talk by declaring that he would not reference any Microsoft products or technologies. The approach differed dramatically from typical COMDEX speeches. It signaled that he had evolved his role from Microsoft's chief marketer to industry visionary. Videos described the limitations of PC computing at the time in offices, schools, and homes and envisioned solutions that he anticipated coming in the not-too-distant future. At the time, few PCs connected to local area networks—much less the Internet. Near the end of his speech, Bill put up a slide that laid out the components of the IAYF vision.[3] It looked like this:

Information at Your Fingertips
- More "personal" personal computers
- Transparent application integration
- Integrated fax, voice, and electronic mail
- Company-wide networks without complexity
- East access to a broad range of information

---

3   Bill Gates's 1990 Comdex Keynote "Information at Your Fingertips" can be accessed at https://youtu.be/tWd8DxLfDek.

The IAYF speech was a call to action for the entire computer industry. Unit sales of PCs had flattened, and Bill pointed the way to renewed growth and prosperity for all. He ended the speech with a call to action, "This is a formidable challenge. No single company is going to be able to do this. It's going to require hardware manufacturers, software developers, and the distribution channels all cooperating to make the vision of information at your fingertips a reality."

Where did this vision come from? I was not in the COMDEX hall that day, and I was not privy to the origin story. Jon Lazarus has pieces of the story. Jon recalls that Steve Ballmer, Bill's right-hand man, came to him with an assignment months ahead of COMDEX. Steve and Bill had been talking, and they agreed that Microsoft needed to present a vision of the PC industry's future. Steve called it a "platform." The platform would consist of core technologies and interoperability standards that would enable seamless interplay between PC hardware, the Windows operating system, the applications that people used, and the networks that gave them access to information.

Bill worked with Steve, Jon, and other Microsoft leaders to transform the IAYF platform idea into words and pictures. Over the next half-decade, they added more layers and nuance to the vision. "Once Bill gave the speech, reality followed," Jon recalls. "We were creating a vision that would attract attention across the industry. People had to respond, and people within Microsoft were tasked with making it come true."

People at Microsoft felt surprised. Bill had not laid out his vision internally before the keynote. But they moved quickly. For instance, the applications group responsible for Excel, Word, and PowerPoint endeavored to incorporate IYAF features in what would become Microsoft Office, the integrated suite of products that ultimately took the market by storm.

The operating systems group responded by developing the concept of "integrated storage." Integrated storage was Bill's idea. Every computer operating system has a file system which is an essential component because it provides a structure for organizing and storing

information. In PCs, we use the visual metaphor of nested file folders containing all kinds of data and information, from documents
and photographs to video clips and software programming code.
Bill wanted to create and embed a flexible database in Windows that
would make it easy to share data and information between applications. For example, email contact information could integrate with
the word processing application or sales-tracking software. I played a
leading role in the earliest steps of developing SQL Server. I ran program management for the OS/2 operating system then, so naturally,
the new project drew me to it.

However, the key figure in our first attempt at creating integrated
storage was Jim Allchin. Jim was a top-notch computer scientist
who had developed object-oriented technology as a PhD candidate
at Georgia Tech and as the CTO at Banyan, a pioneer of network
operating systems. Bill recruited him to overhaul Microsoft's woefully
inadequate LAN Manager network operating system. After Jim joined
Microsoft in early 1990, he killed LAN Manager. He focused instead
on developing technologies that he believed would be essential elements of the Windows NT (for New Technology) operating system,
which, over time, was a huge success.

Windows NT and SQL Server provided essential components of
Microsoft's campaign to become a major player in business computing. To launch the project in 1988, Bill hired Dave Cutler and his
engineering team from Digital Equipment Corp., where they had
designed the company's highly successful VMS operating system. Dave
and the team at Microsoft undertook the arduous task of building a
new operating system. NT was Microsoft's first purely thirty-two-
bit version of Windows that could handle more demanding business
computing tasks. They created both a client-side operating system and
a server version designed to host applications, data, and other services. This work took place in parallel with another, much larger team
that advanced the desktop version of Windows, which later emerged
as Windows 95. Windows NT was a complex project. The initiative
began in 1989, and the first commercial version arrived in mid-1993.

While Dave Cutler's team labored away on the foundation of Windows NT, Jim and his crew focused on a skunkworks project code-named Cairo. At that time, I was working on OS/2 and the alliance with IBM. When Microsoft and IBM went their separate ways in 1991, Microsoft stopped working on OS/2, and I joined the Cairo team.

Cairo focused on next-generation technologies we expected would be added to later generations of NT. Those pieces included a new net-worked directory, networked security technology, a file system called the object file system (OFS), and a new user interface. Ultimately, the directory, security system, and UI folded into Windows, but the engi-neering team broke its pick on the file system.

OFS was Microsoft's first attempt at fulfilling Bill's vision for IAYF and integrated storage. The idea was that data could be stored in one place and used anywhere. Computer users could run simple que-ries for information regardless of that data's location. Further, they could access information stored on server computers as easily as their PC hard drives. A relatively new computer science technique called object-oriented programming made the magic possible.

Since the approach could store structured files and data types as searchable "objects"—discrete elements encapsulated in code—that information could remain searchable within the Windows user inter-face or applications.

As program manager for Cairo, I learned a lot about data manage-ment. At Microsoft, a program manager ensures a software product focuses on customer needs but does not control the engineering team or the overall technology strategy. In this case, Jim Allchin led those things and was my boss.

One of my most profound insights then was the importance of viewing files as data. Much of the world's most important data resides in files, like word-processing documents, spreadsheets, and emails. (Since all our work happened before the internet revolution, we did not need to consider web pages as files.)

Files are complex structures that, in most cases, are only under-stood by their applications. As a result, the data remained trapped within them. Therefore, a key challenge in fulfilling Bill's IAYF vision required cracking open all those files to mine them for nuggets of information about their content.

Integrated storage was a two-part problem. One part involved converting files into searchable objects accessible from anywhere. The other part required a new technique that helped computers understand the content inside those files.

I do not want to get into the gritty details of our struggles with OFS. That would interest only the most obsessive students of the history of data management. Suffice it to say we ran into technical barriers we could not overcome. When Jim Allchin looks back, his takeaway is that Microsoft made everything too complicated. Bill asked for a single system that handled content indexing, queries, and an object approach for data structuring. The system also needed to work across a network of machines despite the slow and error-prone process involved with remote file access. It was too much. "I should have walked away from it, but I didn't," Jim says.

In the end, the grand vision of Cairo did not play out as we had hoped, and OFS was a bust. In 1994, we announced Cairo as an update to Windows NT but could not pull it off. Finally, in 1997, we combined the Cairo development team with the Windows NT team. That was the white flag of surrender. But creating an object-based database within the operating system that all Windows applications shared remained our goal. From the late 1990s into the first decade of the 2000s, teams worked on a succession of integrated-storage technologies.

At different points, we attempted to integrate SQL Server's core technology into Exchange Server and the Outlook email client. But it did not work. Email and calendars, for example, contain semi-structured data that does not follow the tabular structure of SQL databases. We found it impossible to translate semi-structured data into the rigid tables that SQL Server required.

We demonstrated some of our nascent technologies to software developers outside the company but never fulfilled Bill's dream.

Jim recalls the frustrations of working on these projects. Before a critical engineering performance review meeting, Jim bought a statue of a giant metal pig with wings and set it on the conference room table. When the group convened, as Jim recalls, "I told them, 'This is what you're building. It ain't ever going to fly.'" In 2013, Bill told a journalist at *The Register* that this failure was the biggest disappointment during his time at Microsoft. It was mine, as well. Over the years, I started, worked on, or killed several integrated-storage projects. In some cases, I did all three.

While Cairo struggled, the European Organization for Nuclear Research (CERN) pursued a different approach to the IAYF problem. In 1989, Tim Berners-Lee wrote a paper describing the World Wide Web. In 1993, the National Center for Supercomputing Applications (NCSA) introduced the Mosaic graphical web browser, which promised to address many of the problems Microsoft struggled to solve. Unlike Windows, the web stored data in an open format, making it accessible to any application. Shortly after the introduction of Mosaic, Netscape Navigator became the first commercially successful web browser, and Netscape became a Microsoft nemesis. But that is a story for a different book.

Returning to IAYF, the other half of the solution to the problem came from Google with its advances in text-based web search. Google was not the first internet search product. Text search emerged years before Google launched in 1998. A Digital Equipment Corp. (DEC) offering, AltaVista, established an early lead in capability for internet search. AltaVista was the first searchable, full-text database on the web with a simple interface. Later, Google founders Larry Page and Sergey Brin discovered a better way. Their PageRank algorithm delivered superior search results by analyzing the relationships among websites. The process stack-ranked search results based on the number of previous user visits and the importance of the web pages linked to them. It was an early example of the "wisdom of crowds," the catchphrase coined by author James Surowiecki in 2005.

Why was Microsoft unsuccessful with its dream of putting the world of information at our fingertips? Microsoft saw the importance of the web and successfully pivoted in 1995, but Google embraced the open approach when Microsoft leadership did not. The choice allowed Google to succeed in capturing the search market, a leadership position in consumer advertising, and ultimately most of the browser market. Microsoft saw the importance of the web and open standards, but its leadership could not imagine solutions that did not center on the personal computer.

In retrospect, PC-centric integrated storage was never going to succeed. While building the pieces of integrated storage was possible, installing the services and keeping them working was an endeavor only the most prominent and best-funded corporations could undertake. The internet and, eventually, the cloud changed the picture and opened the world of applications and data to companies of all sizes and individuals.

While it took much longer than anticipated, and the journey was very different than what Bill imagined back in 1990, he was right that it would take the entire industry to fulfill the vision of "information at your fingertips." I am pleased that the connected world we live in today—with the web, smartphones, applications, and access to information of all types—very much fulfills Bill's initial IAYF vision.

CHAPTER 3

DATA POWER TO THE PEOPLE

Microsoft was born as a desktop computing software company. As it gained strength, it eventually dominated the PC industry alongside Intel, the leading microprocessor supplier. But by the late 1980s, it became clear to Microsoft's leaders, in particular Bill Gates and David Vaskevitch, that an excellent opportunity existed for growth in business computing. It could benefit data processing for accounting, marketing, HR, manufacturing, and more. At the time, only the larger businesses could afford expensive mainframes and minicomputers and had access to the most powerful and cutting-edge data-processing and reporting capabilities. Medium-size companies made do with networks of PCs and spreadsheets, and most small businesses did not own computers. Many technology companies contributed to democratizing business computing, but Microsoft delivered these new capabilities on a mass scale for all companies. Microsoft gave data power to the people.

A lot of the credit for getting this started at Microsoft belongs to David Vaskevitch. In 1971, while working at the University of

Toronto, he designed and programmed one of the world's first email systems. Later, at the networking company 3Com, he helped Ethernet inventor Bob Metcalfe establish the technology as the industry standard for running wired and wireless networks. In 1986, Microsoft's president at the time, Jon Shirley, lured David to Microsoft. David's first job, oddly, was to build up the company's marketing function. He had never managed marketing before. David worked on several strategic projects in his early years at Microsoft. In my view, the most important thing he did was whisper in Bill's ear about the potential for building a large software business serving the enterprise market. Doing so meant supplying an array of run-the-business software products and services. At first, Bill rejected the idea. As David recalls, Bill told him the company lacked the expertise to build large-scale software systems.

But David did not give up on enterprise computing. He wrote a detailed business plan and began hiring technology and business experts who understood the enterprise market. He realized that to succeed, Microsoft needed its own highly scalable relational database. Finally, Bill gave him the go-ahead to build a database team, and, in parallel, Bill launched the effort to develop Windows NT. Meanwhile, to get a foothold, David and Bill engineered the connection with Sybase and Ashton-Tate to jointly resell Sybase's database, as described in Chapter 1.

A pivotal moment for David and Microsoft's enterprise ambitions came in 1993 when David had the nerve to attend the annual International Workshop on High Performance Transactional Processing. This invitation-only affair occurred yearly at Asilomar, a stunningly beautiful conference center on the California coast. Event attendance was the computing-industry counterpart of being invited to the Vatican—where only the most accomplished computer scientists were welcome. David met and courted people from DEC, IBM, Oracle, and other business computing giants there. Afterward, he hired ten of them. The new team members became the intellectual foundation for Microsoft's foray into enterprise database computing.

One of those hires was Peter Spiro. Peter is a bearded computer scientist with a powerful New Hampshire accent. He spent the previous decade as a database architect for DEC, where he helped engineer Rdb to be the fastest database in the world. At the time, DEC planned on selling its software business. Peter did not like the idea of working for any of the likely acquirers, so he started looking for a more amenable situation. That turned out to be Microsoft and the enterprise initiative. "I was hired by Microsoft to be a change agent," Peter recalls.

When Peter arrived in Redmond, only six engineers worked on SQL Server, led by Ron Soukup. SQL Server was one of several separate database engineering projects underway within Microsoft—each vying for the company's future enterprise database product. The SQL Server team collaborated with Sybase engineers to improve the database. David and Bill had already decided that Microsoft needed to develop its version independently on the foundation provided by Sybase's codebase. While Soukup and his small engineering team did excellent work compared to the hundreds of engineers at Oracle and other database companies, Microsoft needed to grow the team and completely change the architecture and code to compete.

Peter started as one of two vital engineering leads. As the organization grew, it divided into two teams: the relational engine team led by Hal Berenson, with Goetz Graefe as the technical lead, and the storage engine team led by Peter, with David Campbell as the technical lead. Eventually, Peter became the leader of the entire project.

Peter did not race into this. From the beginning, he wanted to create a sustainable engineering culture that could last for decades. At the time, software programmers in the PC industry were often young and inexperienced, and they typically darted from one project to another and sometimes between companies. Peter recruited seasoned programmers and kept them happy and productive for years or even decades. Some of the engineers Peter hired in the mid-1990s are still on the SQL Server development team today. Peter aimed to replicate at Microsoft the engineering culture that thrived at DEC.

Peter's engineering culture centered on the core value that each team member saw themself as a professional committed to quality. That meant they would produce code with as few errors as possible, and when they found problems, they would fix them immediately. They would fail fast and fix fast. There was no expectation that engineers on the SQL Server team would work from sixty to eighty hours a week, which was typical in the industry then. Fifty to fifty-five hours was plenty. Peter did not want people to be exhausted and, therefore, prone to making mistakes.

It was all about quality. At the time, tech journalists thought that Microsoft's first release of a new product would be horrible, and the second major upgrade would be okay. The third version, however, would be a killer that worked smoothly and contained the features most customers wanted. Peter pledged that the first Microsoft version of SQL Server would be the killer, and the engineering team delivered on that promise. Back then, Microsoft software products typically shipped with software bugs that customers reported, and the Microsoft engineers fixed them. Each team would put out a new version of the product a year or so after the original release with bug fixes. When SQL Server 7.0 launched in 1999, it was the first pure Microsoft product, and there was no need for version 7.5.

I wrote earlier about the importance of instilling fundamental values in an organization and want to elaborate on how we implemented that in Microsoft's culture.

Paul Flessner was Peter's counterpart on the business side. He was a CIO for a Fortune 100 company that did serious transaction processing with minicomputers and mainframes. He joined Microsoft a few months after Peter did. At the time, Paul Maritz, one of Microsoft's top executives, spent a lot of time talking to corporate customers. He visited Paul Flessner's company and pitched the capabilities of Microsoft's brand-new Windows NT operating system for servers. Paul Flessner recalls being so impressed with Microsoft's vision that he followed Paul Maritz into the parking lot and told him he wanted to work there. A recruiter from Microsoft called him the following week.

Paul started as a program manager on the SQL Server team. His job was talking to enterprise customers to find out what they wanted in the product. When the project's general manager became ill, Paul was offered the GM job on a probationary basis. He said he would only do it if the job was his. The bosses said yes. He held that job for the next ten years, managing the business to its first $100 million in sales, then $500 million, then $1 billion and beyond.

But before sales charts had a growth curve looking like a hockey stick, Microsoft needed a product. That was Peter's job. An enterprise-scale database has two major components. First, it needs a storage engine, which governs how the system accesses the records on a disk. The other part is the relational engine, which understands the SQL query and translates it into instructions that the storage engine can execute. There was much technical experimentation in the 1970s and 80s. When Peter and the team worked on SQL Server, we embraced the best techniques for designing and building enterprise relational database products. The time was ripe for modernization. Microsoft would reinvent SQL Server using proven architectures and approaches and adding features that met the moment's needs. For example, sophisticated applications required a feature called record locking, and it posed considerable engineering challenges. Meanwhile, Microsoft's well-established competitors grappled with ill-fitting legacy technologies from the past. When the company released Microsoft SQL Server 7.0, nearly all of the relational engine's code and about 80 percent of the storage engine code were new.

Microsoft's database was less expensive and easier to use than those of Oracle, IBM, and the other dominant players. A traditional database requires a large amount of fine-tuning by a corporate customer's database administrators. Those administrators had to configure dozens or even hundreds of settings correctly to get the database to perform well. With SQL Server 7.0, it worked right out of the box and self-optimized as it processed data. Also, SQL Server 7.0 required less maintenance, lowering the "total cost of ownership," a term that

Microsoft popularized. Paul's comment: "We changed the economics of enterprise computing."

Paul recalls that his mom wanted to understand the product he worked so hard on in the early days. "She asked if a database is something people would like to receive as a gift. I told her no," Paul says. "But, looking back, it really was a gift. Our customers wanted to reduce capital and operational costs, and this did that for them."

Peter, Paul, and their teams produced versions of SQL Server for different-sized companies. They gave the database away for free to software developers. They also built a desktop version. That move is a no-brainer now, but it was controversial then. Some engineers considered desktop computers toys and did not believe an enterprise-scale database could run on them. But, through innovative engineering, the team got it done. This effort produced a family of products that served a vast market. "We were running everything from bingo applications on a desktop to nuclear power plants and train systems in Japan," Paul says.

One of the top factors in the rocket-launch success of SQL Server 7.0 was the work that Paul and others did to create an ecosystem of technology partners. Microsoft had successfully run this play with MS-DOS and then with Windows. It was the most important tech ecosystem on the planet. Microsoft's program managers copied and applied the Windows platform/partnership model to other products. Encouraged by the success of Windows, hundreds of software application companies jumped on the Windows NT bandwagon, and a great mix of application vendors, systems integrators, and value-added resellers did the same with SQL Server 7.0.

Paul's most formidable and strategic partner was SAP, the dominant player in run-the-business software applications for enterprises. SAP's CEO back then was Hasso Plattner, a brashly confident German who initially did not have much regard for Microsoft's enterprise database. Later, he changed his mind. Oracle made aggressive moves in the enterprise application business, and Hasso wanted to hobble Oracle any way he could. SAP ported its applications to SQL Server 7.0.

"I don't believe SAP at first really believed in our database, but they wanted some leverage against Oracle, so they went with it," Paul says.

David Vaskevitch tells the backstory: One day, he got a call from Hasso, who was in Seattle and had his teenage daughter with him. Hasso wanted to talk face-to-face. The two men sent Hasso's daughter and David's teenage son off to explore a science museum, and then Hasso and David spent three hours walking the streets together, hashing out the framework of a deal. In that agreement, SAP would release its R/3 application suite on SQL Server within ninety days after Microsoft shipped a database version with record-locking as a feature. No money would change hands, but each company would assign sixteen top engineers to work on the porting project in the other's offices. The deal worked and paid off for both companies.

Paul and Peter remain close friends. Both eventually left Microsoft, but they sometimes invest together in startups—mostly in the social entrepreneurship space. They have advice for people who want to transform industries: "It's three things, and you need all three," says Paul. "First, you need a great team. If you have a great team, you have a chance to build a great product, and if you have a great team and product, you have a chance to build a great business."

Peter agrees but says company culture must support the three elements. "Your culture drives virtually everything you do in engineering," he says. "Do you care about hiring great people? Do you care about performance and quality? Do you care about customers? Do you provide great management and leadership? This list goes on and on. We very consciously thought about all this stuff and more."

When I look back on this era, I am struck by what David, Peter, Paul, and a host of other people at Microsoft accomplished. They disrupted an industry and made the power of data available to everybody. They could do it because Microsoft's desktop products were so successful and profitable, which enabled the company to invest a lot of money and time in developing its enterprise products. Bill gave them plenty of slack. In the early days, he left the engineering teams alone. And later, though he did a lot of reviews, he did not rush the process.

I am often reminded of the early days of Microsoft SQL Server because I built a data center on my property outside Seattle to test new products. That was in 2004. I ran the Windows Server division then and wanted to understand in detail how our products worked. So, I had our building contractor install a raised-floor data center in a room complete with all the usual data center bells and whistles. At one point, I had 11 servers on five racks running all Microsoft's Server products—SQL Server, Exchange, System Center, and SharePoint. It was a crazy thing to do, but I gained firsthand experience with the products my teams built, and thus I could provide better direction.

If having a data center in my home seemed outlandish in 2004, it is nuts today because everything has moved to the cloud. All my servers are gone now, but the data center remains. It is a white elephant from a different time.

PART TWO
THE MODERN DATA STACK

Fast-forward to the incredible progress in data analytics over the past decade. At the center of these advances is the modern data stack, a collection of products and services that work together in the cloud, making gathering, managing, integrating, analyzing, and sharing data easier and cheaper. The modern data stack is the technological foundation of the data economy, and it promises to help people and organizations solve some of the world's most challenging problems.

The modern data stack is an industry-wide solution to data analytics that shares three characteristics:

- It is delivered via software services.
- Leverages the public cloud for scale and low cost.
- Models data for use by a SQL data warehouse.

When envisioning the modern data stack, viewing this technology ecosystem as a closed loop is most helpful.

The modern data stack starts with people and the applications they use in their business. Typically, there are many data sources, including software-as-a-service and enterprise applications, relational databases, third-party datasets, application log data, and information streaming from devices of all types.

Modern Data Stack

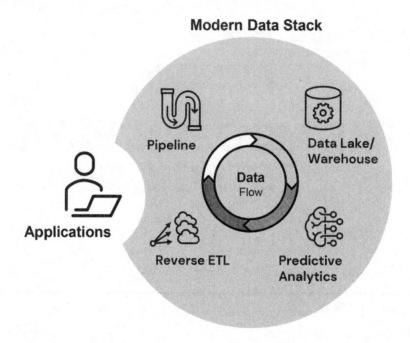

Next, data pipelines transfer the data across networks or the internet and copy it into a centralized location in the cloud. Two popular approaches to managing data in the cloud are data lakes, which can be used to store all types of information, and data warehouses, which use the SQL language to perform queries or transformations on data.

After data flows into a data lake or warehouse, it is transformed from its initial structure for operational applications into an analytic schema used for analysis within a SQL data warehouse.

Data analysts work with a SQL data warehouse using business intelligence (BI) tools such as Tableau, Looker, and Power BI. The combination of BI tools and a cloud data warehouse provides a powerful, modern platform for various analytics applications, including dashboards that managers use to discover actionable insights.

In parallel, data scientists use sophisticated statistical models to analyze the historical systems of record contained in data warehouses and lakes. Developers create these models using various languages,

including Python, Scala, and C++. Machine learning can then interpret data to understand what is happening within a business and make
predictions.

Finally, the data produced by analytics can feed back into the
operational applications used daily by people within a company. The
processes that perform this integration are called Reverse ETL, a reference to the data-ingestion approach called ETL, which stands for
Extract, Transfer, and Load.

The modern data stack is a closed loop because data starts its
journey with people and the applications they use. After processing
through the modern data stack, that information returns to applications and people.

In the upcoming chapters, I'll highlight several datapreneurs and
companies that have made it possible.

CHAPTER 4

DATA MASS-MIGRATES TO THE CLOUD

Seeing how Bill Gates's vision of IAYF continues to influence the computer industry is gratifying. For instance, at Snowflake's annual user conference in June 2022, Benoit Dageville, one of the cofounders, referenced it in his keynote speech. Benoit told the crowd, "Our first innovation pillar is about putting all kinds of data in your data cloud—putting all data at your fingertips."

Snowflake, which I led as CEO from June 2014 until April 2019, became one of the most celebrated cloud software companies. Its founders had grand ambitions from the start. Benoit and cofounder Thierry Cruanes, working with venture capitalist Mike Speiser from Sutter Hill Ventures, launched the company in 2012, intending to develop the first data warehouse designed to run in the cloud. Over time, they created a robust software-as-a-service platform for managing, analyzing, sharing, and selling data. Now they are extending that into a cross-cloud platform for building applications.

By moving data to the cloud and into Snowflake, customers could access unlimited computing and storage resources, support unlimited concurrent users, and pay only for the services they use. Customers no longer needed to set up, provision, and manage their own data centers.

Demand was so strong for Snowflake's benefits that it established itself as the fastest enterprise software company to achieve $1 billion in revenue. It also managed the largest-ever enterprise software IPO, valuing the company at nearly $70 billion on the day it went public in 2020.

Snowflake's technology has evolved to the point where it can now manage all kinds of information, including unstructured data like video, audio, images, and more. It can handle all the core tasks that business analysts, data engineers, and data scientists require. Those high-end scientists and analysts no longer need to pull data out of Snowflake to get their work done. The company automates the loading of streaming data, and its developer environment natively supports applications written in Python, the programming language favored by data scientists.

The modern data stack enables organizations and individuals to store, organize, and analyze all kinds of data, helping users understand how the world works to optimize their strategies and operations to quickly adapt to changing conditions and predict what will likely happen next. Snowflake's data cloud has become an essential piece of the modern data stack for many companies. Snowflake also works closely with partners across the industry that provide products for data loading, business intelligence, data visualization, and analytics.

The world's digital data is mass-migrating from on-premises data centers to the cloud thanks to Snowflake and other modern data stack service providers. It is part of a megatrend of computing moving to the cloud.

A traditional data warehouse, circa 2012, was a SQL database that brought together business data from various sources to enable organizations to ask questions about that data. Conventional data warehouses in organizations' on-premises data centers were expensive to operate. While they excelled at analytics, many companies found that on-premises solutions could not scale to handle the required workloads. Data and work needed to separate into different systems to overcome this challenge, creating data silos as a problematic by-product. Each needed to be maintained and kept in sync with the others.

To make matters worse, as I mentioned earlier, the early data warehouses could not manage semi-structured data. Because web and cloud applications generated prodigious amounts of semi-structured data and insights could be derived from that information, many organizations tried to use a package of open-source software called Apache Hadoop to run queries. Hadoop used a brute force approach to data analytics, distributing the work to clusters of computers that the technology would orchestrate. There were, and still are, many problems with Hadoop. It is difficult to install, maintain, and manage and often requires specialized technical talent. But one of the biggest problems was that most organizations committed to Hadoop used it on premises in their data centers. So, they missed out on the elasticity and ease of cloud management.

Beyond the critical decision to build for the cloud, Benoit, Thierry, and a third technical cofounder, Marcin Zukowski, made several other decisions early on that set the company on the right path.

The first decision separated storage from computing. In the cloud, a business can maintain one reliable copy of all the data collected. Data warehouse software can track that information and make it simultaneously available to any number of computers. The approach circumvents the problems of having multiple copies of the data scattered across various databases, especially since some could contain inconsistent or incorrect information. Separating storage from computing also meant organizations could store their data securely and economically in the public cloud and only pay Snowflake when they used the data. That encouraged organizations to move a large amount of data to the cloud to be managed by Snowflake.

With Snowflake, data management was not the job of an operating system like we had envisioned at Microsoft with our concept of integrated storage. Instead, the cloud handled it all and provided a much better solution.

Snowflake's founders' second major design decision was changing how they used computing resources. The cloud is elastic, allowing the addition of computing resources quickly to complete a particular task. Snowflake's engineers designed their system to assign big computing jobs to a cluster of computers appropriately sized for the task. There was no waiting in line for computing resources. As a result, Snowflake could support an unlimited number of simultaneous users and data queries processed in seconds rather than minutes or hours.

Snowflake's architecture also made data sharing easier. Data managed in Snowflake helped internal business units and external partners share information without making copies. The approach eliminated the data silo problem.

When I arrived for my job interview in May 2014, the company's headquarters housed thirty-three employees in two rooms in a small brick building near the railroad station in San Mateo, California. I met first with Benoit, Thierry, and the rest of the management team. They had no revenue, but after Benoit explained the technology architecture, it was instantly clear to me that if this approach worked, it would profoundly impact and change the world.

After I joined the company, I met with hundreds of potential customers and pitched Snowflake to them. When I described the features and capabilities of Snowflake to seasoned data warehouse professionals, it was typical for somebody at the table to interrupt and say, "I know data warehouses, and what you're saying isn't possible." That was the moment of truth. I would smile and show the slide in my presentation that laid out Snowflake's architecture. We might spend forty minutes or more on that slide alone. In the end, the question I often heard was, "Why aren't other data warehouses designed this way?" My answer was simple, "Those existed before the cloud made it possible."

The founders had solidified their architectural vision when I took the CEO job in June 2014, but it was far from clear then that the product would work. That effort required a lot of innovation by

Snowflake's architects and programmers. It is challenging to make a powerful system easy to use. I knew about making enterprise-ready software, so I helped out. In early 2015, our weekly all-hands meetings changed from a general discussion to an engineering review dedicated to finalizing our product. Progress accelerated, and we released the first commercial version of Snowflake's data warehouse in June 2015.

Developing a successful business model was one of my jobs as CEO. The founders had separated storage from computing, but our solution needed an appropriate pricing scheme. Customers would pay commodity pricing for storage and only pay for the computing resources that they used. Because Snowflake could supply unlimited computing resources nearly instantly, customers could get their tasks done much more quickly and for much lower prices than was possible under traditional computing and pricing models. They flocked to Snowflake.

Another major initiative was to work with the founders and employees to develop and codify the company's values. We started this after the first commercial version of the product shipped. I wrote earlier about my belief that companies achieve long-term success through their values. Adherence to that philosophy should drive *what* they do and *how* they do it.

In the early days at Snowflake, the company's culture and values were informal and based on Benoit, Thierry, and Marcin's egalitarian and collaborative personalities. The company's office had an open floor plan. Everybody sat together. I recall my first day on the job as CEO. Nancy Venezia, the office manager, pointed me to an empty desk among the programmers. I sat down and got to work.

I understood from my years at Microsoft that a growing organization needed a clear mission and the values that supported it. At Snowflake, our mission was to help every organization become data-driven. Over eighteen months, we developed our values like a company builds a software product.

A handful of the early software engineers started defining our values. Later, every department joined in. After much deliberation, we built Snowflake's culture around eight principles: Put Customers First, Integrity Always, Think Big, Be Excellent, Get It Done, Own It, and Embrace Each Other's Differences.

Put Customers First came from Chris Degnan, the company's first salesman and now Snowflake's chief revenue officer. I admire Jeff Bezos and how he instilled the importance of customer-centricity through-out Amazon's culture. Snowflake wholeheartedly embraced that sentiment, too. Since we always put customers first in our daily work, that value deserved the top spot on our list. We lived that value intensely. I remember our dedication to the success of Localytics, one of our early customers. They embedded Snowflake in their mobile analytics platform. If we went down, it would mean they were down—so we made darn sure our service did not go down!

At first, we had seven values, but there was concern in Silicon Valley at the time about increasing diversity. After all, different view-points, backgrounds, and insights help companies make better prod-ucts. Our office manager, Nancy, nicknamed "Snow Mom," came up with our last value, Embrace Each Other's Differences. We added it a year later, and it was my favorite. From the day I joined the company as employee number thirty-four, I felt we had something special in how we treated one another. That came from respecting others' opin-ions and having the desire to listen.

Nancy was employee number seven. She was the company's HR department for the first few years, onboarded new employees, and served as a confidante for the staff. As a Mexican American, she had experienced discrimination. Also, a member of her family was gay. She ensured nobody at Snowflake felt like an outsider. She flew gay pride flags in the office and organized international potluck suppers. When Nancy suggested adding the eighth value, she said it was vital to demonstrate that we welcomed everybody when a new employee arrived. "We wanted to give everybody a place at the table," Nancy recalls. "That was a big part of what it meant to work at Snowflake."

In those early years, the company leaders made some difficult deci-sions and stuck to them. The foremost example was the decision to operate only in the cloud. We believed it was the right thing to do, not just for us but for our customers. At the time, as I mentioned before, many large organizations owned their data centers. They purchased and managed their hardware and software and needed to ensure they had the excess computing capacity to handle peak demand anytime. That meant their servers often sat idle, making their on-premises approach very expensive and inefficient for data processing.

Large enterprises would not even meet with Snowflake in the early days. They committed to on-premises computing and invested significantly in hardware, software, and training. Plus, many did not trust the security provided by cloud computing. In reality, well-run cloud solutions can deliver better protection than the vast majority of on-premises systems.

One of the first large enterprises that granted Snowflake an audience was Goldman Sachs, the Wall Street investment-banking giant. In late 2014, before Snowflake released the first commercial version of the product, Benoit and the sales team made a presentation to a room full of Goldman technology leaders. As Benoit explained the technology, the Goldman people nodded in approval. Then one of the bankers asked when Snowflake would release a version designed to run in a customer's data center. Without a pause, Benoit answered, "I predict that you will move your computing to the cloud before we move our technology on premises."

The Goldman people laughed, but then they realized Benoit was serious. Snowflake was committed wholeheartedly to cloud computing. The meeting broke up awkwardly, but at least one of the Goldman leaders became a believer that day. That was Matt Glickman, who had been at Goldman for twenty-four years. Within days after the meeting, he decided to leave Goldman and join Snowflake.

While the Goldman meeting disappointed Benoit and the rest of us, that was not the end of the story. Several years later, Goldman became a Snowflake customer. Many other large enterprises followed suit. By early 2023, Snowflake had more than 570 Forbes Global 2000 customers.

Another reason Snowflake succeeded was its commitment to the relational database and SQL. For decades, the relational database and SQL were the most effective tools for storing and accessing structured data collected by traditional business applications. Remember how I described the relational database earlier? It organizes data in tables of columns and rows where SQL can query it readily. Because of this model, analysts can use a declarative method for describing data and writing queries. They write a SQL query that specifies *what* they want without having to describe in detail *how* the database management system should retrieve it.

In general, SQL did not assist in cases involving semi-structured data. That type of information typically describes behind-the-scenes actions within an application, a website, or a data center. Every time a person logs on, a recording of the activity and everything they do with the application occurs. This information is usually stored in a semi-structured format. It can be analyzed to improve the application, manage security, personalize experiences, and improve sales in e-commerce applications, among other uses.

Data often resides in nested hierarchies somewhat reminiscent of older database technology. However, semi-structured data does not fit easily into the rows and columns of a SQL database. Instead, it organizes in more flexible ways that computers can still understand. There, it is available for analysis using a variety of data formats. It maintains a textual representation of the data when stored in JSON format.

Snowflake's creators designed the cloud data warehouse to manage semi-structured data more simply and make it more easily searchable. For instance, Snowflake's architecture allows it to access semi-structured and structured data using a single SQL query. Before Snowflake came along, Apache Hadoop managed and analyzed this kind of data, hindered by all the negatives described earlier.

Accommodating unstructured data came later. That data includes documents, pictures, video, and audio content. It also encompasses files associated with a specific application or domain like architecture, genomics, engineering, microelectronics, or aeronautics. Unstructured data is still underutilized. The market research company IDC projects that by 2025, 80 percent of the world's digital data will be unstructured. Snowflake's technology leaders initially did not address unstructured data in their designs. But as the platform matured and they added features, unstructured data was an obvious target.

I think "unstructured data" is a misnomer. These data types are structured—just differently. People understand them, but they are opaque to computers. I prefer to call these data types "complex" rather than "unstructured." If we can manage their complexity, we can more easily combine them with other data types and draw insights from them. And that is what Snowflake and other data platforms like Databricks enable us to do.

Before cloud data products came to terms with complex data, its organization typically manifested as lists of items in file folders stored on premises. In cloud storage systems, Amazon created a powerful service called S3 that stores "blobs" of data within a "bucket." A blob can be any photo, movie, or other files a database uses. S3 supports trillions of blobs, and it is very reliable. Using this foundation of blob storage, organizations added these complex data files to their data lakes, which are systems for storing and managing various data. Snowflake and other modern data stack providers combine data lakes and data warehouses to enable customers to work with all types of data—structured, semi-structured, and complex.

Complex data analysis proves to be ultra-valuable for all sorts of applications. Medical centers, for instance, can run machine learning algorithms to identify patterns in X-rays or MRIs that might indicate the presence of tumors. Retailers can tag photos of clothing items so their e-commerce engines can respond more precisely when customers search for content on a website. Call centers can access recordings of conversations, use natural language processing (NLP) to transcribe them, and then perform sentiment analysis on the transcriptions.

Snowflake's leaders made a sound decision early on by focusing on the widely used SQL language for managing data and writing queries. They gradually opened the aperture to support other languages natively, starting with JavaScript and then Java and Python. These languages are essential for managing and analyzing complex data types. While data scientists use SQL widely, Java and especially Python are favored by data scientists, the gunslingers of the data economy.

Python was named by its creator, Guido Van Rossum, after the 1970s British comedy TV series *Monty Python's Flying Circus*. The only funny thing about Python, though, is its name. Python is a super-flexible programming language that is relatively easy to use. It helps to write web applications, query databases, create data visualization dashboards, model future scenarios, and manage machine learning-based applications.

Snowflake began to support Python natively in 2022. Previously, data scientists typically pulled data out of Snowflake and wrote their queries or did predictive analytics on desktop computers or within a cluster of servers, often running Apache Spark and Databricks. By supporting Python natively, Snowflake enables data scientists to do their work within the Snowflake environment and tap into clusters of computers from there. As a result, the data scientists save time and money.

As I write this, Snowflake is completing its evolution from a cloud data warehouse to a soup-to-nuts manager of all types of data accommodating the most popular programming and querying tools. It has three core businesses: the data warehouse, a data marketplace where customers can easily share, sell, and buy data, and an application-development platform. The first two were in the works when I was CEO. The third came later. In 2022, during the conference where Benoit made the "all data at your fingertips" statement, Snowflake introduced a handful of new capabilities to strengthen application development. For instance, built-in technologies help customers to develop, deploy, and monetize applications within the platform.

The cloud data world continues to change rapidly, and that evolution results in today's clash between Snowflake and Databricks. Snowflake started with the business analyst in mind, focusing on easy-to-use SQL for queries and modeling. Databricks catered to data scientists who use languages like Python to write queries and data analytics algorithms and to transform data. Now both support all types of users. Snowflake expanded its support for programmers and data scientists, and Databricks seeks to address business analysts' needs.

Ali Ghodsi, Matei Zaharia, Andy Konwinski, Arsalan Tavakoli-Shiraji, and others founded Databricks in 2013. Working at the University of California, Berkeley, they were the original creators of Apache Spark. Spark is an open-source software package that makes it easier for data scientists and engineers to manage data processing tasks on clusters of cloud-based computers instead of laptop workstations. Databricks created technology that enabled analysts and programmers to manipulate data using computer languages like Python and Scala. They could also write sophisticated queries using a general-purpose mechanism called DataFrames, or perform predictive analytics using machine learning.

I remember the first time I heard of Databricks. It happened at my first Snowflake board meeting in the summer of 2014. Mike Speiser, the venture capitalist who helped start Snowflake, described Databricks as a company with a high-talent team and promising technology that could eventually emerge as a competitor. That put Databricks on my radar. But, early on, Snowflake and Databricks partnered, and many customers used both products. We aligned so closely that we even discussed a merger at one point. The idea grew from an informal conversation when Ali and I cohosted a party in Las Vegas during Amazon's AWS Reinvent conference. Nothing came of that conversation, though.

Today, Snowflake and Databricks are archrivals. Each company has strengths and weaknesses, and competition benefits customers by forcing rapid innovation. The primary battleground is the data lake, the data management structure accommodating semi-structured and complex data types. It stores data in a proprietary format called FDN, which stands for "flocon de neige," or Snowflake, in French. The loading process creates searchable metadata for SQL to query, making it possible to execute queries in seconds or less. In contrast, Databricks uses an open file format based on a data lake and calls it a "lakehouse." Snowflake has responded to Databricks and now supports an alternative, open data lake format.

Beyond their rivalry, Snowflake and Databricks also compete with other cloud infrastructure providers' data management and platform technologies. Those companies include Amazon AWS, Google Cloud, and Microsoft Azure. I see Snowflake, Databricks, Google, AWS, and Azure as the Big Five data platforms, and the competitive landscape is good news for data-centric organizations. It creates a broad ecosystem of dozens of companies that produce advances in the modern data stack. While we have made much progress, some problems still need solutions. But the Big Five are on it.

Will one company eventually come to dominate this market? I doubt it. Currently, many enterprises use more than one platform, and all these companies have armies of highly competent computer scientists and aggressive salespeople. I do not expect winner-takes-all competition.

That said, when I look back on my five years as CEO of Snowflake and the progress the company has made since I left, I see the profound role that one company can play in advancing technology. Snowflake helped the data economy explode by creating the first effectively limitless database. Thanks to Snowflake's data cloud, there are no ceilings on the volume of data, the number of database tables, or the number of concurrent system users. Snowflake also showed that the relational model could play a critical role in the era of Big Data, where managing structured, semi-structured, and complex data within a single technology environment is essential.

Data clouds like Snowflake's are critical building blocks for emerging AI assistants. The analytic platforms gathering massive amounts of data are rich sources of knowledge that can train machine-learning models. Tapping into this knowledge and opening predictive analytics to a broader set of businesses and organizations is an opportunity that we will see develop over the next few years.

CHAPTER 5

EVERYTHING CONNECTS

In the early days of Snowflake, transferring data from existing business systems into our product presented a challenge. We needed to partner with other technology companies to build a complete solution.

One of our earliest partners was Fivetran, founded by two lifelong friends, George Fraser and Taylor Brown, in Oakland, California. The duo came from Y Combinator, the famous Silicon Valley startup incubator, just a few months after Benoit and Thierry left Oracle to set up shop in a small apartment south of San Francisco.

Fivetran provides the internet's most efficient universal data pipeline, connecting any data source to any destination. Think of Fivetran as being like a plumbing system that reaches rivers, lakes, and aquifers and delivers a steady stream of clean water to businesses. Instead of water, though, it's data.

Combining Fivetran with a data warehouse like Snowflake's or Amazon's Redshift and a data visualization tool like Looker, we had the needed pieces for a modern data stack. After joining Snowflake as CEO and getting to know George and Taylor, I liked Fivetran so much that I offered to buy the company. George wisely turned down the offer. Now, I am an investor and a member of its board of directors.

Before the cloud data revolution, storing data was expensive and problematic. In the mainframe era, data resided on big, monolithic computers. When networking took off, organizations kept their business information on servers using a combination of files and databases. Typically, data was "owned" and managed by an enterprise's business units or functional organizations. However, the individual data silos created by that approach made analyzing information across the organization difficult.

This situation did not improve when enterprises started using data warehouses in their on-premises data centers. Much of the data loading required complex processes that few people understood. It took thousands of individual mappings, resulting in a web of entanglement. We call that procedure "extract, transform, and load", or ETL. Because data warehouses remained expensive, organizations often kept only their most immediately relevant data there. Even when those systems loaded a subset of the data, with some pre-aggregated, it could take a day to load. Also, business logic typically applied to the data during the ETL process makes debugging difficult.

The cloud revolution turned the data world upside down. Suddenly, storing and managing large amounts of information within data warehouses was cheap. But it was still hard to move data into these data warehouses. Many cloud customers wrote custom programs in languages like Python for this task.

Fivetran's powerful technology addressed the problem by efficiently moving data from business systems into a cloud data warehouse. Organizations could transfer it more easily with the automated process. Because of the emergence of economical elastic cloud warehouses, businesses could store all their data in the cloud. Thanks to Fivetran, they could manage that data more flexibly in response to changing business needs. Fivetran led the transition from ETL to ELT—Extract, Load & Transform, which means the cloud data warehouse does all the work of transforming and preparing the data for analytics.

George and Taylor have a fascinating shared history. George grew up in the northern suburbs of New York City, and Taylor grew up in Colorado. For generations, their families have gathered in the summer at neighboring lake cabins in Wisconsin. The two guys were like brothers. Later, George attended CMU, got a PhD in neurobiology at Pitt, and worked as a cognitive scientist at a biotech company. He loved computer programming and dreamed of building data-management technology for scientists. Taylor went to Amherst College and pursued fine arts. He later studied design strategy with a focus on the technology user experience.

Their careers took them to San Francisco, where they reconnected. George worked alone in his apartment on data analysis tools for scientists. Taylor also had the itch to start a company, so they joined forces. They named their company Fivetran as a play on Fortran, one of the leading programming languages for mainframe computers, which had become the lingua franca for scientists. Eventually, they discovered that there was insufficient demand within the scientific community for the tools they worked on, so they shifted to the data analytics market.

They focused on data integration and ways to move data from existing operational systems to data warehouses where it is more easily managed and analyzed. At the time, organizations put a lot of time and resources into manually formatting and moving data. George and Taylor saw an opportunity to develop prefabricated connectors for loading structured and semi-structured data into cloud data warehouses. "We were party crashers. We did not know the domain, but that actually helped us. We benefited from our ignorance," recalls George. We studied systems, talked to many customers, and came up with a different solution. While we met some resistance initially, we eventually came up with a set of features that people liked."

The first apps that Fivetran supported were Salesforce, HubSpot, and Zendesk, some of the most popular applications that cutting-edge Silicon Valley startups were using.

Today, Fivetran sells hundreds of zero-configuration and zero-maintenance connectors that continuously adapt as organizations change their data schemas and the Application Programming Interfaces (APIs) they use for accessing data from applications. Fivetran continuously synchronizes data from source applications to needed destinations, enabling analysts to work with the freshest data. Because all customer data runs through a Fivetran pipeline, centralized bug fixes can improve reliability and accuracy in the loading process. Fivetran manages some initial transformations within data warehouses using an open-source technology called Data Build Tool (DBT).

Fivetran creates DBT packages for a wide variety of data sources. After the datasets load into the cloud data warehouse, the DBT packages act on the data, producing ready-to-query schemas that data scientists can customize to handle more exotic tasks. "The most important thing we do is hide complexity from the user," George says.

Fivetran's design is critical to its success. Before Fivetran, most data management and analytics software products required highly technical people, including data engineers and data scientists. Taylor and George aimed to create a product that could help many organizations and people, not just the techies. "We agreed that our product had to be ridiculously simple," Taylor recalls. They tested early prototypes with Taylor's mom, a nontechnical person, to see if she could set up a connector. She could!

While Fivetran provides some transformation features, the focus now and for the foreseeable future is data movement. The company will continuously add more connectors, ensuring that customers have access to all relevant information. Fivetran also builds solutions for large enterprises to synchronize their mission-critical operational databases with the cloud data lake or data warehouse of their choice.

My interactions with George and Taylor are a healthy two-way street. When I met them, I knew a lot about databases but not so much about data movement. One of the top lessons they taught me is the importance of *idempotency* in data integration. In computer science, a system is idempotent if it always converges on the same result when you run the same operation multiple times. Idempotency is a critical concept I applied to other elements of data management at Snowflake and beyond it.

In the data integration process, replication and movement can create problems because the procedure must run smoothly without introducing gaps or flaws in the data. If data pipelines are non-idempotent, they break, and a highly paid person must fix them. In contrast, a properly functioning idempotent pipeline can repair itself when, for example, a previously unavailable data source comes back online. With Fivetran technology, if an operation fails in the middle of the loading process, it corrects automatically without the need for expensive interventions.

I gave George and Taylor a lot of encouragement and reinforced they remained on the right track. I also coached them on the many issues of effectively running a company and helped them build their enterprise business. Ultimately, though, steadfast determination fueled Fivetran's goal of creating the world's best data pipeline "that just works" for companies of all sizes.

Whether custom-built in Python or created using a service like Fivetran, data pipelines are essential plumbing that plays a critical role in data infrastructure by bringing the data to cloud platforms. These data pipelines provide the supply chain for today's analytic applications. They also perform a crucial function in developing machine intelligence, like the human body's central nervous system brings sensory information to the brain. Tomorrow's AI assistants will train using data lakes and warehouses connected by data pipelines.

CHAPTER 6

VISUALIZING DATA

I distinctly remember the first time I understood the power of the spreadsheet. I moved from Michigan to California and was working for ROLM, where I was a team member that developed software on Data General minicomputers and IBM PCs to configure ROLM PBX systems. I must have shown some sort of spark because the bosses invited me to a retreat for high-potential employees. It occurred at Pajaro Dunes, the seaside resort on Monterey Bay famous for hosting many of Apple's engineering and management retreats.

Junior executives like me split into teams and competed to complete a challenge. Before the retreat, organizers gave us a few pieces of paper with a business scenario, a P&L, and a set of market conditions. They also gave us a sample data set. I worked day and night before the event creating a Lotus 1–2-3 spreadsheet that could model the business and allow us to ask "what-if" questions.

On the first night of the gathering, I stayed up late to load the dataset into the spreadsheet on an IBM PC. Our retreat was before

Windows, so we looked at columns and rows of glowing numbers and letters on a dark background. The organizers asked the teams, "How do you run the business within these data parameters? How many people do you hire, and in what roles?"

I was the token techie on our team, but the spreadsheet quickly revealed that company executives running the offsite had specified an unlimited-demand scenario. That means theoretical customers bought as many widgets as we could produce and sell. We developed a proposal calling for hiring as many salespeople as possible quickly. We called our proposal the "Win or Beach" plan because we figured if we were wrong, we'd quickly run out of money and be out of the game so we could go on the beach and enjoy ourselves. It was a no-lose situation. But we won!

You do not run into unlimited-demand scenarios very often in the real world. But I saw it happening in 2018 at Snowflake. Our demand data, available using our Looker BI tools operating on Snowflake's internal data warehouse, told us we should hire as many salespeople as possible. And we did. That concerned some board members, but it turned out the right decision. Snowflake's sales took off like a bottle rocket.

Why was the spreadsheet so powerful in the early days of PC computing? First, spreadsheets were easy to use and had a wide range of uses. Anybody with a rudimentary understanding of PC software could set one up to manage anything from a garden club mailing list to a small business accounting system. More sophisticated technology users could run simulations and what-if scenarios. Second, you could see and manipulate the raw data. Unlike a database, you did not need to write a query. What you saw was what you got, and it did not change unless you changed it. Over time, spreadsheets became even more powerful. With the arrival of the graphical user interface, tables transformed into visual charts and graphics with the press of a button. Lotus engineers invented the pivot table, which made it possible to recognize patterns easily and perform calculations on a subset of the data in a spreadsheet, including sums, averages, and other statistics.

Combining all this goodness, we got the first killer app in the history of personal computing—something so compelling that millions of people and businesses bought PCs.

Dan Bricklin got things going. While a student at Harvard Business School, he observed his professor's frustration when needing to erase and replace data constantly on a lecture-hall blackboard. He thought, why not create data tables digitally to update them easily? He and his business partner, Bob Frankston, produced VisiCalc spreadsheet software to do just that. When first available in 1979, VisiCalc could run on Apple II personal computers. The success of VisiCalc was one of the main reasons IBM developed its personal computer.

Microsoft released its first version of the Excel spreadsheet in 1985. It was the first graphical spreadsheet running on Windows, and by the mid-1990s, Excel led the market. In the early days of Excel, Microsoft ran the company on it. It was the basis of our financial management system. I consider Excel the most significant piece of software Microsoft ever made. It has the distinction of being used by millions of people to keep lists of just about anything while also being used by financial service companies to make some of the most sophisticated business calculations on Earth.

Spreadsheets remain hugely popular. More than 750 million people use Excel regularly, according to Microsoft. I expect that spreadsheets will thrive for decades to come.

Spreadsheets have limitations, though. Compared to a SQL database, querying a spreadsheet is awkward. A tab within a spreadsheet is essentially one table, whereas a database is a collection of many tables connected by a relational structure. However, because spreadsheets and databases are interoperable, spreadsheet tables can import into databases, and database queries can export into spreadsheets for analysis. I think of a spreadsheet as a small and fast ferryboat and a database as a huge and powerful ocean liner. Each serves a different but vital purpose.

One bridge between old-world PC spreadsheets and today's mass data analytics phenomenon was the business intelligence (BI)

technology of the 1990s and early 2000s. These tools combine querying, reporting, online analytical processing, and dashboards for viewing
and monitoring data. Like many other advances in the early decades
of computing, the concept of business intelligence started at IBM. In
1958, IBM computer scientist, Hans Peter Luhn, published a landmark
article titled *A Business Intelligence System*. He explored approaches to
draw insights quickly and easily from vast data sets to help make the
best decisions. For decades, BI tools beyond simple reporting capabilities remained costly and only available to large enterprises with a lot
of money and expertise. But a lot of smaller companies needed these
capabilities, too. Spreadsheets are great for working with relatively
small amounts of data. When organizations deal with bigger data, they
need a bigger gun.

That gun came from Tableau Software, launched in 2003 by
Christian Chabot, Pat Hanrahan, and Chris Stolte, data visualization
experts at Stanford University. Tableau's technology enabled users to
pull data from various sources, mix and match it, and generate graphical representations of the data. The technology made it easy for nontechnical people to manipulate and understand their data. Industry
observers labeled this "self-service BI." Tableau was superior to Excel
in several ways. Tableau made it easy to integrate data, prepare ad
hoc reports, and create dashboards with compelling visuals. For years,
Tableau remained the dominant BI product running on Windows. At
the time, Microsoft did not have its own BI tools.

The guy who brought BI to Microsoft was Amir Netz. He was running R&D for Panorama Software Systems, which he had cofounded
in Israel. Microsoft bought the company for its online analytics processing technology in 1996 and incorporated it into Microsoft SQL
Server.

Amir is a technology prodigy. He learned programming on mainframes at a nearby university during his early teens. His parents broke
the family budget to give him an Apple II computer and a printer for
his bar mitzvah. That is how he came upon VisiCalc, the first spreadsheet for PCs. It blew his mind. At age 16, he wrote an entire suite

of productivity software. However, his high school nearly kicked him out because he neglected his studies. After university studies and military service, his big break came when software entrepreneur Rony Ross hired him as the initial employee of Panorama Software in 1993. There, Amir developed the first Israeli BI application for PCs. He wrote the first version in ten days, and the product became a huge success in Israel. When he and Rony made plans for a marketing trip to the United States, they reached out to Corey Salka, who was then the Microsoft program manager for Excel. They asked for a meeting, and wisely, Corey said yes.

What happened next was like a technology-industry fairy tale. The two met with Corey, demoed their current technology, and laid out their vision for what would come next. Corey was so impressed that he immediately introduced them to David Vaskevitch, who ran the division. As Amir recalls, David offered to buy the company on the spot, asking Rony to name her price. Stunned, she suggested $18 million, which seemed like a lot of money when she said it. It turned out to be a magnificent bargain for Microsoft. The technology was the foundation for SQL Server Analysis Services and Power BI, two of the most successful data analytics technologies ever produced.

Rony promised that Panorama would deliver the newer-technology version of the software in just six weeks. When she and Amir left the meeting, they immediately called their colleagues in Israel and canceled all vacations. After Amir returned to Israel, he lived and worked in the office for weeks. The developer team managed to meet the deadline. The BI application was speedy despite PC technology's limitations at the time. Amir invented a powerful way to reduce the amount of prep work required to make PC-based BI systems operate efficiently. Amir recalls that their beta version was full of glitches, yet, miraculously, none surfaced when he demonstrated it to Microsoft technologists visiting him in Israel. He and the team passed muster, and the acquisition went through.

Amir became Microsoft's leading BI thought leader, delivering many technology improvements to Excel and SQL Server. His most significant contribution as an employee started in 2008. Amir helped lead the shift that eventually produced Power BI and moved the company's data analytics technologies to the cloud.

Pre-cloud, as more and more data became available for business analysts, system performance deteriorated. The SQL databases of that era choked on data. Amir saw that speeding query response required systems to store more data in memory. The problem was memory capacity was limited. How could we resolve this dilemma?

His cognitive breakthrough came after a long flight from Seattle to visit family in Israel. During the flight, he read an article in a technology trade publication where the author referenced the nonuniformity of data within a database. Some values require more frequent use than others. Amir realized that this insight was the key to solving his puzzle. After he arrived in Israel, he stayed up all night working on an algorithm that would dramatically improve the way data is stored in memory, making it smaller and much faster to query. This invention became the basis for the Picasso project, the in-memory processing engine that would be the foundation for Power BI, SQL Server Analysis Services, and the multidimensional cube functionality in Excel. "We called it the Engine of the Devil because it was so fast," Amir recalls.

I ran Microsoft's Server and Tools Division when I first met Amir. When he and his colleagues on the Picasso project team decided their processing engine was ready for prime time, they arranged a meeting with me. I remember the day. We were in my conference room in Building 42. Amir and other team members ran their demo and explained what happened in the background. The engine was processing data at an incredible rate. When used with Excel, it allowed users to expand capacity from 64,000 rows in a sheet to 100 million rows. I was stunned and excited. The approach could revolutionize the data analytics business. Excel users could put a giant corporation's sales data in a single worksheet on a PC and slice and dice it to their heart's content. Afterward, I emailed Amir and the Picasso team, saying how impressed I was with their work. "Once in a very rare while, I have the privilege of learning about work being done that's world-changing," I wrote. "I was blown away. This is amazing work. The implications are mind-boggling."

The combination of Amir's query processing engine and Excel made for a great demo, so we arranged for Amir to demonstrate the technology at many tech-industry conferences. He was intelligent, engaging, and excellent at demos. At one point, I invited him to Microsoft's annual sales engineering gathering at the Washington Convention Center, where he showed the technology to 5,000 people. The audience included technical salespeople who helped customers understand Microsoft technology and build business solutions. They loved Amir's demo. I was on stage with Amir. As he wrapped up and began to head off stage, I grabbed his shoulder and said, "There's one more thing." On the spot, I promoted him to Microsoft Distinguished Engineer while the crowd gave him a standing ovation. It was an incredible moment for both of us.

In my twenty-three years at Microsoft, I worked with many smart and innovative people, but technologists like Amir motivated me to go to work every day. They were solving some of the most complex computer science problems of the day and simultaneously developing products that could address the widest audiences at affordable prices. I left Microsoft about a year later, in 2011, but that day on the stage with Amir has lingered with me ever since then.

The way Amir and his team architected the technology underlying Power BI allowed Microsoft products to support large numbers of users and crunch massive amounts of data with little memory. As a result, Microsoft could offer Power BI as a service at a low price and incorporate it into the Microsoft Office pricing structure. The combination of a great product and the strength of the Microsoft salesforce drove incredible growth in Power BI usage.

Earlier, I talked about the importance of Tableau Software in taking the power of BI to the personal computer—and inspiring the folks at Microsoft to develop BI tools. But Tableau was a Windows desktop program, and the company was late to invest in the cloud. That created an opportunity. A key pioneer in cloud data visualization, Looker Data Sciences, emerged in 2011, one year before Benoit and Thierry launched Snowflake. Lloyd Tabb and Ben Porterfield founded Looker, and CEO Frank Bien led it. The company created a simple data modeling language, LookML, that enables business analysts and executives to find, explore, and analyze data without knowing SQL. Looker was well suited to serve as a front end to cloud data warehouses.

Lloyd was the technical genius in the group. He was a top engineer and architect for web browsers at Netscape through its incredible early run. Later Lloyd was among Mozilla.org's founders and continued the development of its open-source browser after AOL bought Netscape. After leaving Netscape, he cofounded a series of startups. Lloyd and Ben launched Looker in Santa Cruz, a laid-back surfer town on the California coast, and brought on Frank a few months later.

At the time, organizations began using massive data sets to make all sorts of decisions. Big tech outfits, including Yahoo and Google, started using Apache Hadoop and Map Reduce to analyze data, but the technology required specialized skills to set up and run queries. Meanwhile, Tableau made it easier for regular people to query large data sets and visualize the results by loading the data into a visual workbook for analysis. Unfortunately, the process siloed data. Various databases employed different schemas and definitions of things, creating a data Tower of Babel. Frank recalls that database administrators and analysis would have "data brawls" about defining and labeling business information within companies. It could get ugly. Lloyd's innovation was an analytics platform in the cloud that made it easy to work with cloud data warehouses like Redshift and Snowflake. LookML, essentially, was a universal language for describing and explaining data. "Suddenly, people stopped arguing about data," says Frank. "No more Tower of Babel."

Tableau also moved to the cloud. Tableau and Looker served different purposes. Tableau was best for pulling data from multiple sources and assembling quick-and-dirty dashboards. Looker could only connect to one data warehouse at a time but provided everyone across the organization with a consistent view of the data. Google later bought Looker and folded it into the Google Cloud Platform. Salesforce purchased Tableau.

When I look back on the period when Snowflake, Tableau, Looker, and Fivetran came on strong, that is when the concept of the modern data stack came together. The technologies capitalized on the cloud data revolution and helped medium-size companies move their applications and data to the cloud. Companies benefit from Fivetran because it simplifies pulling data into the cloud. Snowflake allows them to manage and share various data sets and types. Tableau, Looker, and Power BI provide tools to fetch and analyze data.

While CEO at Snowflake, I recognized that success came from partnerships with other companies that provide pieces of the modern data stack. Through a partnership, we could approach potential customers with a complete solution delivered as a cloud service rather than one piece of a puzzle.

Early on, we hired a business development leader, Walter Aldana, to identify the most important companies in the field and forge strong partnerships with them. Walter knew that Amazon, Microsoft, and Google would try to kill us once they figured out our value proposition. Each company had its own cloud data warehouse, making us competitors and partners. Walter started his partnership initiative with *them*. He convinced some of the most influential companies in techdom that destroying us was not advantageous for them. Walter successfully argued that we could help them to build their infrastructure businesses.

Walter developed partnerships with data visualization companies, too. Looker came first. The founders were anxious to tie up with us, and we needed them. "Before we teamed with them, only data scientists could understand Snowflake," Walter recalls. "With Looker, companies could visualize more data across more users in a more seamless way than they ever could before." In the early days, Snowflake's initial sales calls featured live demos involving Looker using their corporate data. "With Looker, we could wow customers," Walter says.

Tableau was standoffish at first. It was a well-established company and did not see much use in working with a startup it rarely witnessed in the marketplace. In 2015, Walter called or emailed them around fifty times, asking them to build a connector to Snowflake. No dice. Then he went through the back door to a former university friend, who introduced him to one of Tableau's key technology decision-makers, Adam Selipsky, who now runs AWS for Amazon. Walter knew we probably had just one chance to make an impression, so he enlisted me, the founders, and our heads of sales and marketing. We all participated in the meeting. The pile-on showed Adam just how important a partnership would be for us. But the key deciding factor was the Snowflake technology. He was impressed. Tableau ended up building one of the best third-party connectors to Snowflake.

A fantastic array of data notebook, BI, visualization, and analysis tools are available today. This category will get even more interesting with the introduction of AI assistants into some data tools that serve as portals and help business users access data. Foundation models and, ultimately, AI assistants will enable us to use English and other spoken and written languages to analyze and visualize data. AI promises to make data analysis accessible to more people than ever before.

PART THREE

NEW DATA TECHNOLOGIES

Most of what you have read so far is history. It happened, and while the significance of some events and innovations remains open for debate, there is no undoing it. The remainder of this book is about pathways into the future and new approaches to data management and analytics that could change the course of the data economy. It could also impact every part of society.

What you read next is an informed view of exciting developments in the data industry and speculation about where it will go. It is also a tour of several small companies I am working with that have the potential to help shape the future.

CHAPTER 7

A NEW TYPE OF OPERATIONAL DATABASE

Most of the applications that people use every day do not focus on the past. They work with up-to-the-minute information. The specialized databases used by these applications are called *operational* databases. They are the backbones of applications and data center operations. When you buy an airline ticket or order an item from an online retailer, an operational database captures that data. The operational database can inform you about the order's status and whether flights remain on time. On the other hand, if you want to look at shipments you have received in the past or flights already flown, that information is likely coming from a historical analytic database.

During the 1980s and most of the 1990s, a single SQL database handled both operational and analytical tasks. As applications grew larger and larger, databases could not scale to meet emerging needs. So, databases divided the workload for either operational or analytic use. Data pipelines connect the operational databases to data warehouses.

Operational systems focus on fast and efficient support for transactions. Using the airline ticket example, an airline needs your full legal name, birth date, and credit card to reserve a spot on a flight. The ticket reservation and credit card charge take place through a single transaction, so a traveler gets their ticket after credit card approval. If a flyer reserves or changes a seat, a transaction holds the chosen seat and releases another if needed.

Transactions are equally crucial for data warehouses, but because they work with historical information, data warehouses work with batches of data. With a batch-processing system, many changes update at the same time. Snowflake's platform can search through petabytes of historical information and efficiently load multiple terabytes of data every hour. However, you would never use the Snowflake data warehouse as the backbone of an airline reservation system. Because of these limitations, Snowflake recently introduced new support to handle operational database requirements better. However, that is not the same as a database optimized for operations that perform thousands of transactions per second.

Operational databases are a big business. They are the foundation upon which Larry Ellison built his Oracle empire. Oracle's database and other products like Microsoft's SQL Server use SQL and work with rows and columns within tables. The approach is ideal for highly structured work like placing an online order, booking an airline ticket, or moving money between bank accounts. In these cases, a given type of transaction always updates a well-defined set of columns. Banking requires a very structured environment, making fixed SQL tables perfect for the task.

But many modern operational applications do not fit this pattern and have much more dynamic data storage requirements. An early example of this was email. Email messages cannot easily fit into tables with fixed columns. One person or a hundred can receive an email message and carbon copy (CC) or blind copy (BCC) even more. That email can also contain one or more attachments.

When I worked on integrated storage at Microsoft, we tried to force-fit SQL Server as the email data store within Outlook and Exchange server. Every attempt failed because we tried to use a database optimized for structured data to store semi-structured information. We tried to fit a square peg into a round hole.

Semi-structured is a natural way to express the data entered in the web browser forms used by most operational applications. In today's world, applications that produce semi-structured data abound. In addition to the many applications that use browser forms, social media applications like Twitter and Facebook and chat applications like WhatsApp and Facebook Messenger naturally store their data in a semi-structured format.

Since 2008, Evan Weaver has focused on a quest to build a modern operational database capable of handling semi-structured data with enormous transaction volumes. Evan is a cofounder of Fauna, and my conversations with him sparked my interest in the company.

His quest started after he joined Twitter as employee number fifteen. At the time, Twitter was nothing like today's global phenomenon. However, usage began to surge after its showcase at the South by Southwest Interactive conference in 2007. Before Evan joined Twitter, he had been a member of the Ruby on Rails programming community, which developed a framework for building dynamic web applications. He volunteered to help Twitter's founders improve the application's performance. Instead of accepting Evan's help on a volunteer basis, Twitter's then-CEO, Jack Dorsey, hired him as a full-time member of a new engineering team. At that time, no off-the-shelf database technology could address the performance and scalability Twitter needed as usage soared. The app faced hardware constraints, so the underlying time line system that stored and delivered the tweets had to be highly efficient. But the time line system was also brittle, so it was hard to operate at scale. The engineering team tried to adapt a handful of SQL databases, such as MySQL and PostgreSQL, as well as the NoSQL databases

Mongo and Cassandra, to help solve the problem. Nothing worked to their satisfaction.

To solve the performance challenges, Evan and his colleagues built novel solutions to scale critical data processing in the time line system and databases. It was incredibly work-intensive. In the end, Twitter could access massive computing resources that addressed the performance problems. Other web giants, including Google and Facebook, invested heavily in engineering resources to create purpose-built databases to run their enormous operations. But less-wealthy outfits continued to struggle.

Evan rose through Twitter and ultimately became director of infrastructure. After four years, he left to continue the search for the holy grail for web application scalability. The same year Snowflake and Fivetran started in 2012, Evan and Matt Freels formed a consulting company called Fauna Research that offered its data and scalability expertise to gaming and social media startups. Matt was a key member of Evan's engineering team, although they initially met through their shared interest in Ruby on Rails.

Those consulting engagements helped them develop a vision for a general solution that many web operators could use to deal with the wicked problem of database provisioning and scaling. In 2016, they launched Fauna to develop a flexible, developer-friendly transactional database delivered as a scalable and secure cloud API.

Evan said, "People don't work on databases out of love for the existing solutions. They work on them out of rage. They're just so poorly fit to contemporary problems. But what if you created a system that's designed for what people want to do now? That's what we're doing at Fauna."

Matt, now Fauna's chief architect, seeks to simplify the role of website and app developers. "It used to be a rite of passage for a developer—figuring out how to configure a cloud provider and your database for your application," he noted. "Now engineers don't have to worry about the data. They just build their applications. Fauna reduces the amount of time they have to spend dealing with databases from weeks to seconds."

I am fascinated with this technology—so much so that I signed on to be chairman of the board.

The founders chose the name Fauna—which means all animal life in a particular habitat—because they believe a successful database should provide variety and flexibility for application developers. Fauna's product is a semi-structured relational database delivered as a serverless cloud service. In simpler terms, that means Fauna supports and runs customer databases. It is built from the ground up to store semi-structured data using transactions that span different parts of the globe with blazing speed and accuracy. Global support for operational databases that synchronize transactions across multiple data centers will be increasingly important. What excites me about Fauna is its unique approach to delivering the highest level of transaction consistency with minimal overhead.

Data's location can create challenges for operational systems and transactional databases. Data of all types permeate the modern world. Some comes from software applications on desktop PCs or smartphones, and data centers sometimes generate and store other data types collected from applications. Even more data comes from Internet of Things (IoT) devices. Networks of automated sensors and cameras can monitor what's happening around us and store that data worldwide.

In all these cases, location matters. Regardless of how fast we make our networks, data cannot move more quickly than the speed of light. Moving it over long distances takes time, particularly around the globe. For example, if you live in New York, it takes about one-tenth of a second for data to move between your smartphone and a server in San Francisco. The farther away you are from the server, the longer it takes.

Data used in large computing applications typically resides around the globe. In operational systems, data is particularly time-sensitive. A few-tenths-of-a-second delay may not be a problem for smartphone and computer users, but latency can create difficulties for computing systems.

In addition to performance issues, compliance plays a major role in where data is stored. Critical applications require consistent data storage in more than a single data center or cloud region. These applications must survive major failures, including catastrophes like hurricanes, floods, or the total loss of a data center because of a fire. Privacy and government regulations often provide further restrictions, requiring that the data be stored within Europe or even within a specific country. For these reasons, the location of data used by applications matters. In today's world, it makes sense to use a database that provides the flexibility to locate data anywhere. Wherever an end-user resides, data centers around the globe provide fast access to current information.

Business continuity requirements are not new. Banks have dealt with these issues for decades. But customers expected 24/7 access to their applications and data when the world moved online. As a result, the need for multiple data centers and global access is now an essential requirement for many applications.

Often, organizations use two geographically separated data centers to provide business continuity. One was primary, and the other provided a backup. All data transactions occur in the primary data center's database, and a synchronous replication system keeps the backup data in sync. Transactions cannot complete until the geographically separated databases "commit" data and successfully store it to ensure consistency. With synchronous replication, distance matters a whole lot. To keep the data transmission speeds fast enough to achieve consistency, the data centers must be reasonably close together, typically no farther than one hundred miles apart. That is why many New York banks have a primary data center in Manhattan, and the secondary is close by in New Jersey. However, placing data centers just miles apart may not be sufficient to deal with some catastrophes. This approach was problematic when Hurricane Sandy hit and severely impacted both cities.

Fortunately, there is a different approach called a globally distributed database. Google Spanner, the first successful commercial implementation of this approach, was released on Google Cloud in 2017. It was first described in a 2012 paper, "Spanner: Google's Globally Distributed Database."[4] Spanner is a SQL database that—as the name implies—spans multiple data centers wherever they reside. With Spanner, transactions rely on precise time synchronization to commit across various data centers.

Google keeps everything in sync by using a Google feature called TrueTime that ensures that computers within a Google data center have a very accurate internal clock, typically set directly from an atomic clock or GPS.

This approach works well for Google but is challenging for many organizations. Setting a server clock is tricky since many customers use imprecise internet timekeeping services. As a result, server clocks in different data centers can be out of sync by a half second or even more. A half second is an eternity to a database, making fast multi–data center transactions impossible.

Unlike Google Spanner and similar products, Fauna offers a global database that does not require clock synchronization. It can efficiently support transactions spanning a continent or the globe and deploy in any cloud data center worldwide.

Instead of using the time to synchronize the servers, Fauna's approach uses a sequence number. Each transaction is ordered by the system using a technique inspired by a technology called Calvin. Calvin is a serialized transaction scheduling and data-replication algorithm that uses a deterministic ordering guarantee to assure consistency. Serialized transactions provide the highest level of transaction consistency. Dr. Daniel Abadi invented Calvin when he was a professor of computer science at Yale University, with the help of his fantastic research team.

---

4 James C. Corbett, et al. "Spanner: Google's Globally Distributed Database." Google, Inc., 2012. https://static.googleusercontent.com/media/research.google .com/en//archive/spanner-osdi2012.pdf.

Anybody can read Abadi's paper, "Calvin: Fast Distributed Transactions for Partitioned Database Systems."[5] But the devil is in the details, so building a transactional database with similar technology is exceptionally difficult. Fauna got a lot of help from Abadi, one of the company's advisers. As a result, Fauna was the first industrial implementation of deterministic transaction technology based on Calvin. It scales nearly linearly on a cluster of commodity servers and has no single point of failure.

As a result of tremendous work by a group of brilliant people, Fauna's database provides high data consistency, is cost-effective to run, and can easily be placed anywhere.

I cannot overstate how critical this advance in operational database technology is for organizations. Highly accurate data synchronization is important now but will be essential in the future. A couple of decades from now, we will have autonomous machines all around us. Examples include drones delivering goods, driverless cars and trucks, and robots in our homes and offices. Transactions will enable these new machines to interact with the business systems in the world around them. These machines will have to "talk" to one another. Without that capability, they could crash into each other—and us, too! Devices will handle some of that talking, but much will happen in edge servers located close to the vehicles for performance reasons but orchestrated from afar.

The Fauna technology is a relational database, but instead of using SQL as the primary tool for organizing and querying data, Fauna's database offers a language called Fauna Query Language (FQL X), designed to work with semi-structured data. Like SQL databases, Fauna supports relational operations. But FQL X is a better fit for modern languages like JavaScript. In addition, FQL X supports business logic, which is integrated directly into the Fauna database. Incorporating

5   Daniel Abadi, Calvin. "Fast Distributed Transactions for Partitioned Database Systems." Communications of the ACM, 2012. http://cs.yale.edu/homes/thomson /publications/calvin-sigmod12.pdf.

transaction logic into the database makes high-performance, globally distributed transactions possible.

The primary customers for Fauna are cloud application developers and website developers. Like Snowflake, Fauna is a service, so once a developer creates an account, it handles all the database configuration details and operations for them. "This database unlocks the productivity of modern developers," said Eric Berg, Fauna's CEO. "Development teams can move much faster. They spend less time on infrastructure, and that lets them focus on solving business problems."

I have known Eric since he worked for me at Microsoft as a program manager, and we have kept in touch over the years. After leaving Microsoft, he became the chief product officer at Okta, a successful access-management company, where he worked for nearly a decade. Evan and Matt did a great job of laying the technical foundation for the company, and it was time for a professional manager to take the helm. When Eric left Okta, I happily recruited him as Fauna's CEO in 2020, and he has done a marvelous job growing and maturing the company and building a fantastic team.

While I am a huge fan of SQL and focus primarily on data analytics, two things happened during my time at Snowflake that drove home the need for a modern operational database *not* based on SQL.

Building the Snowflake user interface was an important requirement for Snowflake. Like most modern applications, Snowflake's UI uses HTML and JavaScript. It turns out that JavaScript and SQL are a poor match. JavaScript naturally supports semi-structured data. JSON, translated as "JavaScript Object Notation," is the most common textual format for storing semi-structured data. Working within the rigid confines of SQL makes it hard and unnatural to call SQL from JavaScript. Snowflake uses SQL for everything, so the JavaScript developers building Snowflake's UI complained about how difficult this was.

My second learning came from Benoit Dageville, Snowflake's cofounder and current president. The Snowflake platform uses an operational database and "key-value" store called FoundationDB.

Key-value stores serve as an elementary type of database that accommodates semi-structured data.

Benoit and cofounder Thierry Cruanes considered using a SQL database as the transaction store for Snowflake. However, Benoit realized that the fixed structure of a SQL database was too limiting for a constantly changing service. They ultimately chose FoundationDB instead because it could efficiently support the massive number of transactions per second that Snowflake required. It also had the flexibility that Benoit desired.

SQL will remain important for some operational applications where the rigid, structured tables that SQL provides are appropriate. However, many applications require more flexibility, and we need a new generation of operational databases for semi-structured data that can handle the high-transaction-volume applications of the future. Fauna is one of the first of these.

I look forward to seeing new applications built on the next wave of operational databases.

CHAPTER 8

UNLOCKING COMPLEX DATA

I wrote in Chapter 4 that "unstructured data" is a misleading term for describing video, still images, audio, music, books, and various text communications. While people easily comprehend these data types, computers struggle with them. Machine learning and foundation models will change that.

It makes much more sense to call these data types "complex" rather than "unstructured." Complex data is a collection of many different, unique formats. Now, for the first time, we can train computers to understand these other formats, much as kids learn to read in elementary school and, as they progress, gain proficiency in understanding more complicated topics. As we develop a new generation of applications, a computer's ability to handle and understand complex data types is vital. Once they can do this, we open a new world of applications that can extract information from these data sources, potentially combining them with other data types to reveal new insights.

The story of complex data begins for me with Bill Gates and his vision of "information at your fingertips." Bill and others at Microsoft understood that putting all the world's digital information at the people's beck and call means that data must be discoverable by a computer. At that time, it was very early for video and audio on PCs. We thought mainly about image and text-based files. Of course, the applications that created and used text-based and image files understood the format of these files, but in general, the contents remained opaque.

The Cairo project's developers tried to make complex files searchable, and a critical task in my role as program manager was supporting their efforts to do it. But, at that time, we were years away from having the machine learning mechanisms needed for the computer to understand the information within those files and make it available in response to queries.

Today, digital documents, photos, and videos are everywhere, creating a veritable tsunami of complex data. We have many new uses for it.

Decoding video, for example, has become critically important. There is a lot of focus on building drones, electric vehicles, and other robots that can operate autonomously. These systems use video cameras to understand the world around them. That requires the computer to analyze the video stream and understand what it sees. Processing this complex data is a particularly challenging problem because the robot may need to react quickly to what it sees. That requires an immediate understanding of what is happening. For example, a moving autonomous vehicle must know what to do if its camera detects a child running into the street ahead. These types of situations require processing complex data in real time.

New generations of foundation models dramatically accelerate our progress and open new worlds for previously impossible applications. Machine-learning algorithms made it easier for computers to identify information contained in video, audio, and still images to make it available. But we still have a long way to go.

Neural networks and machine learning are essential for handling complex data. Merriam-Webster defines artificial intelligence (AI) as "the capability of a machine to imitate intelligent human behavior." Computer scientists John McCarthy, Marvin Minsky, and two other colleagues coined the term in a proposal they wrote for a workshop at Dartmouth College in 1956. That event made artificial intelligence a scientific field. The dictionary definition talks about machines that "imitate human behavior." However, that does not mean a computer understands what it does or why. We will discuss the difference between performing a task and understanding its meaning in Chapter 11.

During college in the early 1980s, I took an AI class that primarily focused on creating "expert systems." These programs explored human expertise in a particular domain and created a set of rules for the computer to follow to make rudimentary decisions based on those rules. The rules explicitly define the conditions for an action. Expert systems satisfied the basic definition of AI but hardly demonstrated intelligence. For decades, AI proceeded along various paths, each of which made incremental progress.

The pace of AI advancement increased over the last decade because of machine learning. Computer systems can use algorithms and models to acquire knowledge and adapt without explicit instructions.

Machine learning models draw insights from data and use it to predict what will happen in the future based on what happened previously. The models use that knowledge to craft better strategies or perform more effective actions. The models encoded in algorithms speed up with exposure to additional data.

Aggregating data from various sources creates models that use statistical probabilities to make predictions. Traditionally, creating a model is a multistep process. First, data scientists huddle with business leaders or analysts to identify a business system or process problem. Then, the data scientist develops a new model or chooses among preexisting models to find one that fits the business problem. The scientist feeds the model sample data and adjusts the model if necessary. Finally, a scientist can train the model using massive amounts of existing data.

There are several ways of training machine-learning algorithms, but three are the most popular. With *supervised machine learning*, a data scientist provides labeled examples of data to the AI system. A second approach, *unsupervised machine learning,* employs models that use unlabeled datasets for training. The algorithm learns from interacting with the data and uses that knowledge to identify trends or patterns in new data it receives. A third type of training is *reinforcement machine learning*. The algorithm gets a signal whenever it performs a successful action. It learns by trial and error. Over time, training algorithms use statistical probability to predict what is happening with the data, and the AI system can act on that knowledge.

In a simple example, supervised machine learning can identify verbs within a document through a process called "part-of-speech tagging." With supervised ML, people take thousands and thousands of example sentences and manually tag the verbs within them, identifying and labeling them for the computer as verbs. They then train an ML algorithm using those tagged examples. Afterward, when the algorithm encounters a new sentence, it readily identifies which words are verbs. The accuracy of this prediction is based mainly on the amount of training data provided to the model.

Machine learning applies in many scenarios today because we are learning to improve the accuracy of statistical predictions. Some of the best results achieved today use foundation models created by processing vast amounts of data from across the internet to build highly accurate models. In many cases, these models provide results that match or even exceed the accuracy of experts within a given field. Therefore, having more and higher quality data is critical.

A radiologist attends many years of school to learn to read an X-ray or MRI and diagnose a problem. They will review thousands of scans throughout their career and learn from that experience. After learning from many example diagnoses, an ML algorithm can also identify medical problems. Unlike people, though, a ML algorithm can review millions of X-rays rapidly. With enough examples to train from, the algorithm can outperform an expert in the field.

People naturally learn to perform this type of pattern matching from an early age. The process happens through brain chemistry. Our brains contain billions of neurons that store information and relate new information to what we already know. The neurons form connections that create a biological model of how things work so we can interact with the world more successfully.

Since computers only understand numbers, ML models represent information using numerical data structures created during training. When provided with new information, the ML algorithm presents its understanding of the situation and the recommendation for action in the form of a new series of numbers.

I am involved with several companies taking on the complex-data challenge. While their use cases vary, each business applies ML to achieve its desired results. Three early-stage companies doing this are Pinecone, Docugami, and Collagia.

Let's start with Pinecone. The company's founder and CEO, Edo Liberty, brought Pinecone his years of experience in machine learning and advanced algorithm development at Yahoo and Amazon. He was the research director and head of the AI lab for Amazon Web Services. He and his team dealt with some of Amazon's most vexing machine learning challenges, like recommendations and searches.

Pinecone developed a specialized kind of database called a *vector database*, making it easier to build high-performance applications using complex data. Vector databases help applications search text, images, video, and other challenging data types. This kind of data processing assists with everything from identifying new molecules for life-saving drugs to finding patterns of abuse in financial systems.

A vector is a numeric array that represents examples of complex data. Vectors help perform *similarity searches*, meaning computers try to identify things that are not identical to one another but share many traits. A vector database can store, search for, and analyze vectors.

Building vector databases and systems for using them is extremely difficult and typically demands PhD brainpower. Amazon, Google, and other hyper-scale businesses can afford to hire and pay the teams of talented people required to do this work, but most companies can't touch it. That is why Edo started Pinecone in 2019. He chose the name Pinecone because he liked the image of a pinecone with all those seed scales pointing in different directions, like vectors. The company provides a vector database service that other business use in cloud applications or to build applications. The resulting applications are either sold or honed to serve internal functions. Pinecone's goal is to democratize the management of complex data.

Edo loves his job. "The exciting thing about complex data is that it's complex," he says. "You're connecting real life in all its complexity to technology, and you're building systems that can create immense value."

Pinecone first took on the immense task of interpreting text to gain insights from email threads, product support transcripts, and Slack collaborations. The company optimized its vector database to perform a *semantic search*. A *lexical search* looks for an exact word or phrase, but a semantic search can interpret a query's meaning to improve results. The technology uses natural language processing to understand the semantics of sentences in context and stores those models as vectors in a database.

Handling images or videos—especially streamed video—presents a much more significant challenge. How a computer views and makes sense of images is similar to how humans do it. With humans, light comes through our corneas which filter the light and project images on the retina. There, photosensitive cells register light intensity and color, and the retina sends that detail to the brain's visual cortex through electrical pulses. The visual cortex translates those signals into a form the temporal lobe can interpret. There, the brain compares the new information with our memories and known concepts, allowing cognition.

Computers capture visual images in various ways, including using cameras and scanners, and converts them into digital formats. When presented with a new photo or video, a machine learning program creates vectors that describe the image or video content. Advanced geometric algorithms help a vector database to compare new vectors with those previously stored to identify patterns quickly.

For example, a machine learning algorithm converts a picture of a horse into vectors based on its training from many earlier images. To convert a numeric vector into the word "horse," the vector database compares that new vector to known vectors associated with the word "horse."

Pinecone's technology plays a critical role in this process. When customers process information contained in images and other complex data types into vectors, Pinecone's technology organizes the vectors into a database and makes the information searchable for data analytics or real-time applications. The process helps machines to spot new patterns and features, even without specific training to do it.

Until recently, organizations needed data scientists and engineers to build vector databases and develop applications. Now, vector databases and vector search serve business analysts and executives because the background complexity remains hidden.

While Pinecone's technology is a platform for application development, the other two companies I will tell you about build applications for specific use cases involving complex data.

Docugami's CEO and cofounder, Jean Paoli, is the person who put the x in Microsoft's Office file format names, including .docx, .pptx, and .xlsx. A rebel within Microsoft, he was not only one of the creators of the extensible markup language (XML) file format, which revolutionized document representation, but he also catalyzed a massive shift at Microsoft to adopt open file formats and open-source technology.

You may recall that some of Microsoft's rivals banded together in the early 2000s to try to derail Microsoft Office, which was the most popular desktop productivity package worldwide. They created a less-functional Office-type suite called OpenOffice and lobbied governments and large corporations to adopt it because it used an open file format.

Microsoft's competitors aimed to lock Office out of the government market. They were starting to make headway, but Jean stepped into the breach. Jean joined Microsoft in the mid-90s and immediately tried to convince Microsoft's leaders to adopt XML as the basis for their open document standard. At first, Jean faced resistance from the Office team. But after Bill Gates and Steve Ballmer supported him, everybody else eventually stepped in line and got busy rewriting the Office applications with XML as a key ingredient. Once that task was completed, Jean's job was easy. Government technology leaders wanted an open document standard without giving up Office, and once Office changed to using .docx and other open XML formats, there was no need for a change.

Today, Jean has an automated way to translate documents into formats that computers can readily identify. "I have spent my career deciphering documents and making them understandable," said Jean. "What we're doing now is we're transforming documents into data."

Jean grew up in Lebanon, where multiple cultures and languages intermixed. His childhood languages were French and Arabic, and he loved the written word. At age ten, his inventor spirit emerged after his parents gave him an electronic abacus with illuminated beads. Inspired, he persuaded them to allow him to build a larger (sixteen rows instead of four) electronic abacus with a frame made from the headboard of a bed.

Jean had no access to personal computers during childhood, but when he continued his education in France at age 18, he fell in love with computer science. After obtaining his master's degree, he went to work for startups that emerged out of Inria, the French national research institute for digital science and technology. In the mid-1990s, when Bill Gates pivoted Microsoft to the internet, the company sent recruiters out to find people with unique technical expertise. Recruiters found Jean at a Worldwide Web Consortium (W3C) conference and hired him, practically on the spot, to work on the Internet Explorer web browser.

One key ingredient of early web browsers was Standard Generalized Markup Language (SGML), a format enabling sharing of machine-readable documents for government, law, and industry. Jean saw that SGML was not well-suited for web pages, so he and others within W3C began work on XML, which became a global technology standard in 1998. XML was a breakthrough technology for the internet era because it made it possible to represent documents as data in a text format that could be stored, searched, and shared. XML files could be read and understood by both humans and machines. That technology became a foundation for e-commerce and social networking.

Jean and Docugami are working to fulfill the vision of making information that people create more accessible to machines. The folks at Docugami created a software-as-a-service application that runs in the cloud and automates extracting the context and semantics of words, sentences, and paragraphs in documents. The service converts that information into usable data to automate reports or business processes. Docugami's solution is excellent for evaluating contracts, insurance policies, and other long-form documents. Over time, Docugami has the potential to change the way people work together across organizations to create business agreements.

Jean and his colleagues gave the company and product the name Docugami because of its similarity to origami, the Japanese tradition of paper folding. Expert origami artists take flat sheets of paper and meticulously fold them to create beautiful paper sculptures. Docugami's technology uses the XML data model to transform documents into highly structured data that both people and machines can understand.

Here is how it works. Imagine a scenario where a business must evaluate and organize its written contracts. In the past, someone inherited the tedious task of sorting through a confusing mix of printed and hand-signed PDFs and digitally signed Word documents. The Docugami application's simple interface makes that task easy. With the help of Docugami, an employee can import and select documents to evaluate with a few mouse clicks.

In the background, the software reads each document. It uses multiple techniques to identify the most important categories of information, subdivides the content into useful "chunks" of information, and then evaluates and reorganizes similar written material in the same way. It turns a set of complex documents into machine-readable business objects.

When Docugami completes the initial processing, the user can select chunks of information from their documents and turn that into semantic content used for a report or a new document. It can also create a new business process. Under the covers, the Docugami technology has transformed every document with all of its complexity into a hierarchical, semantically tagged version of XML, which the company calls DocuGami Markup Language, or DGML. It also groups sets of similar documents using the same schema. Think of it this way: traditional document formatting addresses only visual appearance, while Docugami's capability handles the information within a document.

Today, the company uses several sophisticated techniques to make the system work as it does. Docugami creates its own machine learning models and combines these with other machine learning and foundation models. This infrastructure transforms a document, including scanned images, into an XML data model. Docugami turns business documents into data, and the technology is flexible, so as foundation models improve, Docugami can take advantage of this directly. I will be exploring foundation models in more detail in Chapter 11.

There is a lot of concern today about the role of artificial intelligence in society and the workplace. The fear is that AI will take our jobs and leave many humans behind, unable to make a decent living. The work done by Jean and his colleagues at Docugami shows how AI and people can collaborate and work together to make people more effective. AI can do things that are very repetitive for humans and take over routine tasks. That lets us be more creative, innovative, and focused on interactions with others. By working as an assistant,

AI can help make individuals and all sorts of organizations more productive, effective, and creative.

While Docugami focuses on mining documents for knowledge and improving their quality, another company called Collagia focuses on interpreting still images and interweaving them with related information in other databases. Collagia's name is a takeoff on the word *collage*, a curated presentation of visual elements.

The idea emerged from an internal incubation project at Nike headed by Rooney (Roo) Armande, now Collagia's CEO. While Roo worked as Nike's visualization and business intelligence project manager, she pitched the idea of creating an internal tool called VisLab to combine and analyze image and business data for the company's footwear, merchandising, and manufacturing units. She and one of her collaborators, Andrew Hanson, recognized the incredible potential to create a product, so they and another partner, Matt Green, launched Collagia in 2018 to bring the technology to market.

Today, retailers, ad agencies, industrial parts manufacturers, construction suppliers, medical care providers, and others capture and store vast numbers of images used for catalogs and other sales and marketing programs.

Digital asset management systems (DAMS) retain this information for individual items, including minimal metadata like file names or creation dates.

Computers know very little about the contents of the images or how they connect to the core functions of the business. Collagia bridges that gap with its visual data engine. The technology marries image data with business data in a way that helps even nontechnical people to ask computers questions. "Our goal is to democratize image data to make it available for everyone—whether they're business analysts, poets, or musicians," says Roo. "They shouldn't have to depend on data scientists and engineers to get their work done."

The company uses image and pattern recognition machine learning to enrich the metadata associated with each image. Their engine connects with pictures stored in the DAMS and automatically identifies features in each image. Then, it organizes the information based on learned taxonomies and creates easily searchable catalogs. Meanwhile, the engine imports business data from data warehouses and other storage places, maps it to relevant images, and makes both kinds of information available through an easy-to-use query tool. The goal is to help people discover correlations between visual information and business data.

Say an online retailer wants to understand how images of products and product features affect sales and profitability. Do people want to see a pair of expensive climbing pants modeled and photographed in an outdoor setting or in a studio where they can see the cut and texture of the fabric better? "This is about putting the right product in front of the right customer at the right time," said Roo. "This will help consumer companies to increase sales by building and marketing the products that customers want."

Roo reached out to me when she saw that I had left Snowflake. Nike was a relatively early Snowflake customer, and she realized that Collagia's visual data engine would need to interoperate seamlessly with data warehouses. I meet regularly with Roo and give her technical and business advice. She gives me a window into a corner of the tech universe that few people understand.

In today's world, video, images, and text are ubiquitous and rich information sources. Managing and analyzing this complex data is a new frontier. Machine learning and foundation models are taming that frontier, and the business world is beginning to wake up to this opportunity. AI assistants built using machine learning and foundation models containing complex data types will change creative and business processes. It will also increase our efficiency and enable new capabilities, some of which we never dreamed possible.

PART FOUR

A MODEL-DRIVEN WORLD

A major technological shift will transform the business landscape, economy, and society in the coming years. Recent advancements in machine learning bring a new generation of computer models, called foundation models, which radically change how we think about computers. These new models respond directly to commands written in everyday language. For example, applications like Stable Diffusion and DALL-E 2 can generate unique artwork on command. ChatGPT is barely a month old as I write this book, yet it already impacts society and raises new questions about student learning and testing approaches.

Organizations of all types, including governments, can develop elaborate digital models defining their business, operations, laws, rules, and cultures. These models can improve continuously with comparison to real-world data. With the right information, they adapt to shifting conditions and guide us toward the future. In other words, a model-driven world drives evolution.

I remember the first time I witnessed the potential of computer modeling. I worked at ROLM in the mid-1980s. My boss, Paul Anderson (still a close friend), introduced me to simple programs created to help business leaders design or model effective business processes. The concept was great, but the technology was not yet available to make it work. However, the experience had a profound effect on my thinking. It became clear to me that executable digital models could perform actions that could replace traditional coding techniques. Over time, I realized that the general modeling concept applies to any process.

At the most basic level, models represent how we believe entities and processes work in the physical world or how we want them to work. They have a long history. As civilization developed over the centuries, people began making designs—or models—of the things they planned on building. That included everything from sketches of sculptures on Mesopotamian stone tablets to Leonardo DaVinci's drawings of flying machines to dioramas replicating small cities.

In the twentieth century, detailed and sophisticated models could describe almost everything built or manufactured. The technique is commonplace in engineering, chemistry, architecture, design, and other fields that create physical objects. For example, a skyscraper's digital model can contain all its structural, mechanical, electrical, HVAC, and plumbing details. Computer-aided design tools have made the product-design process much more efficient, and 3D printing can turn a digital model into a physical object with little more than a press of a button.

A computer model is a digital representation of the real world. By using scientific algorithms, models simulate how things work. That is why people sometimes call them "digital twins." However, these digital twins are never identical to the real thing. While they can be incredibly useful, models are only as good as the assumptions that go into them. If those assumptions are well grounded, the model can be quite accurate. For example, a mathematical description of the airflow across an airplane wing can help engineers create more effective designs. Today, physical or social systems with many variables, such as global weather or the US economy, are difficult to predict with models. But eventually, machine learning can help us understand these complex processes, too.

As our computing techniques and data volumes advance, we can accurately model the world around us—not just buildings, machines, and products—but organizations, business processes, human language, art, laws, and policy. In addition, we can model potential future situations, run simulations, and use them to improve how our businesses and social systems work.

In 2011, Marc Andreessen, Netscape founder and current Andreessen Horowitz venture capitalist, famously said that "software is eating the world." That is undoubtedly true. But today's software-defined world leaves a lot to be desired. In the next ten years, I predict that models will eat software.

CHAPTER 9

A NEW LANGUAGE FOR MODELING THE PHYSICAL WORLD

When people look back on the history of computing, they typically see new hardware or killer applications as significant advancements for industry, the economy, and society. That's true, but programming languages—the collections of letters, numbers, and symbols we use to communicate with computers—are equally important. What would a mainframe computer be without Fortran, the language developed by IBM in the 1950s for scientific and engineering applications? What would the PC be without C and, later, C++, the foundation languages for Windows? And how would the internet have exploded without HTML, Java, and JavaScript?

Computer languages are the foundations for new products. For instance, Snowflake's original data warehouse required a combination of C++ for the processing engine and Java for the query engine.

I did my first coding at the University of Michigan. We used the ALGOL programming language, mainly used by universities for

instruction. The computer center at Michigan housed an Amdahl
mainframe. As with IBM mainframes, we stored programs on 3¼- by
7⅜-inch pieces of thick paper called punch cards. Coding in those
days involved the painstaking process of storing digital information
by poking holes in the cards. A program could finally run when
a card reader attached to the mainframe loaded those instructions
into memory. As students in those days, we carried around piles of
punch cards, often hundreds of them. Inevitably, we dropped them
and watched in horror as they scattered across tables and floors, typically late at night, just before the project was due. It was hell to put
them back in order.

Unlike some of my classmates, I was not a programming whiz kid.
I spent a lot of time learning how to do it, but I thought of programming more as a tool to aid my career. While at Microsoft, I understood
how vital programming and programming languages are.

C++, Java, C#, and JavaScript are some of the killer programming languages of the computing era for building applications. They
make everything from accounting software to the International Space
Station systems possible. Today, a new class of programming languages
supports the data economy.

Decades from now, when people look back on the emergence
of data as a critical element in the global economy, a handful of
computer languages will stand out. SQL, of course, will be one of
them. Other languages, including Python and R, are ultra-popular
today with data scientists and other visionaries in the data universe.
Computer languages are effective for turning ingenious mathematical
algorithms into beneficial code, and each has its strengths. Python,
for instance, is very flexible and has vast libraries containing algorithms and predefined functions, making it easier for data scientists
to develop and refine their programs quickly. R primarily aids statis-
ticians. Engineers and economists favor another language, MATLAB.
But each has weaknesses, including a significant challenge they have
in common. While they are fantastic programming languages, they
face the so-called "two-language problem."

The two-language problem plagued computer scientists for a long time. Modern data-oriented programming languages provide a layer of abstraction that helps data scientists and business analysts write analytics programs and queries quickly, try them out, and make improvements. That process is effective when an analyst operates in a small-scale mode. However, programs written in high-level interpreted languages like Python and R cannot run efficiently. Therefore, a data scientist developing a Python algorithm or application must endure the labor-intensive process of rewriting it using a computer language like C++. The effort requires highly skilled and expensive software coders to rewrite the algorithm and efficiently run it on a large scale.

While programmers had deep experience with the go-to languages for science and engineering, they felt frustrated. They wanted an efficient language to handle large amounts of data, accommodate various programming tasks, and remain applicable for scientific computing projects. The language also needed to help translate algorithms into code quickly and run applications efficiently on a laptop or in massive clusters of computers without rewriting it into a lower-level language.

Fortunately, in 2012 computer scientists at the Massachusetts Institute of Technology (MIT) created a no-compromise, open-source programming language called Julia. Julia handles the two-language problem with ease and makes coding fast and easy.

Viral Shah, a cofounder of the Julia project and CEO of JuliaHub, a company set up to commercialize Julia, says the language "takes the drudgery out of data science." Julia is ten times more productive than C++. In other words, Julia lets developers write one-tenth of the code required by C++ and run computer instructions almost as quickly. Julia also helps scientists and analysts create more accurate computer models, allowing them to make better simulations and predict what will happen in the future more accurately. Also, because Julia is so easy to use, it can be used by more people. As Viral put it, "Data is everywhere, but it's worthless unless you can make use of it. This is all about making sense of the world."

Viral's journey toward Julia began in 2007 after he joined a company called Interactive Supercomputing in Waltham, Massachusetts. Viral had obtained his PhD in scientific computing from the University of California, Santa Barbara and joined the effort as a senior scientist.

The goal of Interactive Computing, later acquired by Microsoft, was to make parallel computing widely available to millions of scientists and engineers through its Star-P development platform using MATLAB as the programming language. The idea was to enable its users to concentrate on high-level analysis rather than getting lost in the weeds of machine-level programming. Over time, it became clear that MATLAB was not a great fit for the task, so Viral and his colleagues, Alan Edelman, Stefan Karpinski, and Jeff Bezanson, began working on Julia. They established it as an open-source project to help hundreds of other programmers worldwide.

Writing a brand-new programming language is a huge undertaking. Adding to that challenge, Viral moved to Bengaluru, India's version of Silicon Valley. He spent the next several years helping the country take full advantage of the digital revolution. His role involved designing policies and technology behind the government's *Aadhaar* project, a national digital identity card and payment system. Then he coauthored a book, *Rebooting India*, with Nandan Nilekani. Nandan led the Aadhaar project and was a founder of the Indian software services giant Infosys. The book focused on ways technology could help India become a global economic powerhouse and improve the lives of the country's 1.2 billion citizens.

Viral helped develop Julia while working remotely. The breakthrough moment for the group came in 2012 when they launched a website with a blog post, "Why We Created Julia,"[6] explaining their mission. They wanted a programming language with all the capabilities of their favorites but none of the drawbacks. "We are greedy. We want more," they wrote.

---

6   Bezanson, Karpinski, Shah, and Edeman. "Why We Created Julia." Julialang.org, February 14, 2012. https://julialang.org/blog/2012/02/why-we-created-julia/.

In 2015, the Julia creators and early contributors Keno Fischer and Deepak Vinchhi established Julia Computing to provide support, training, and consulting services. They chose Viral as CEO based on his work in India. The company developed JuliaHub, a cloud platform for developing and deploying Julia-based programs. The company eventually adopted JuliaHub as its name.

The JuliaHub platform lets people scale computing tasks from a single node to hundreds or thousands of CPUs and supports machine learning targeted to specific industries. A service called *JuliaSim* enables physical modeling and simulation of devices, while *Cedar* assists in electronic circuit design. The *Pumas* application, built in partnership with Pumas.ai, provides an integrated modeling and simulation platform to accelerate drug development for the pharmaceutical industry.

A big reveal came in 2021 when the company announced it had raised $24 million. That was when I joined the JuliaHub board of directors.

I first learned about Viral and Julia through my work with RelationalAI, discussed further in Chapter 10, where I was an investor and board member. RelationalAI chose Julia as the core programming language for its application development platform. I considered that a risky bet because, at the time, Julia was not a widely used programming language. The RelationalAI team had some specific reasons for choosing Julia, insisting that it was unparalleled for data applications. So, I took a closer look. The closer I looked, the more impressed I became. The big benefit of Julia when designing computer systems and applications is its efficiency. It makes digital model building easier, quicker, and cheaper. However, Julia's most crucial attribute is how it enables *automatic differentiation*. This capability is essential to the emerging model-driven world.

In the physical world, mechanics, electronics, and materials are critical elements of a model. In the computer world, the counterparts are data types, data structures, algorithms, and the mathematics that underlie them.

Airline operators carefully maintain aircraft engines to avoid unexpected downtime or failure but must also be mindful of preventive maintenance costs. Today, a *digital twin* identifies the perfect balance by combining a computer model with real-world data to make informed predictions. Using the airplane example, engineers create a software model to duplicate the engine's behavior. In parallel, the flying plane captures real-time telemetry and sensor data. By comparing the recorded data with the model's expected values, airlines can accurately predict problems before they occur.

Derivatives, using calculus, make automatic differentiation possible by measuring the rate of change in a function. This behind-the-scenes math lets models detect unexpected changes in telemetry data, thus signaling the potential for future failure.

Automatic differentiation is a fundamental tool engineers use to design and optimize physical products based on a model. Traditionally, engineers relied on pre-built libraries to perform automatic differentiation, and if that was insufficient, they turned to C++ programmers to complete the task. The Julia community made automatic differentiation a first-class feature in the language. Programs written in Julia can perform automatic differentiation across applications and help organizations to design different components of their products using coordinated software tools.

Modern devices and their components, such as electric vehicle batteries, can benefit from telemetry data analysis enabled by automatic differentiation. Digital twins can detect problems before components fail, ensuring the replacement of troubled parts before they impact service.

Julia helps build models of the physical world so engineers can better understand functional nuances and ensure a product's components run optimally. Engineers and scientists start by developing a hypothesis based on limited knowledge. They then build a software model to test the theory. They may run hundreds or thousands of simulations, constantly improving the model. Ultimately, they produce an ultra-accurate model that simulates how the device will operate in the real world.

While it is still early for Julia, data scientists already use it in many applications. For example, investment management giant BlackRock uses Julia for time-series analytics, a statistical method for analyzing past data within a specific period to forecast the future. Julia also helps the Federal Reserve Bank of New York to model the US economy. Scientists worldwide depend on Julia to help address climate change using next-generation modeling.

While automatic differentiation has existed for a long time, Julia enables this capability to be easily and consistently applied across the different engineering tools required to build a modern product. When I look at today's world, I see opportunities to apply automatic differentiation everywhere. Chapter 11 will describe how automatic differentiation is critical for neural networks that underlie foundation models.

With more than 45 million downloads, Julia's adoption is enormous. Today, 10,000 companies and 1,500 universities worldwide use it. The JuliaHub founders expect to see Julia among the top twenty most-used languages for new applications this year and among the top ten in the next five years.

CHAPTER 10

A NEW PLATFORM FOR DATA APPLICATIONS

Businesses use many applications to manage their operations, keep track of their finances, collaborate with others, and make decisions about everything. Software can help determine how many widgets to make to how to plan for an approaching hurricane. All these applications tap into data stored in various databases, often in the cloud. In most cases, applications also store newly created data there. In accounting, people review expenses to verify accuracy and due dates and then issue payments. Other people track sales data and go through the painstaking process of changing SKU prices in product inventory when their analysis tells them it is time to do so. That is a lot of manual labor!

A new generation of computer programs called *data applications* (also known as *intelligent applications*) can automate that tiresome process in the years ahead.

To understand data applications, it helps to compare them with typical apps we use daily. Applications allow users to send emails, schedule appointments, and order products online. These everyday tools interact with people, working for us and with us. However, data apps are different because they respond to data, not people.

Most data apps built today use the modern data stack. Changes in data activate them, and they use machine learning and other predictive analytic services to act independently. Those actions could include issuing an alert, sending a marketing email, initiating a business process, or purchasing a block of stock based on changes in market conditions. The data tells the application what to do rather than people doing it, although humans still make the rules.

Data applications enable a new level of automation of routine tasks that frees people to do higher-order work like inventing new things, analyzing data, strategizing, developing new goods and services, improving business processes, and collaborating with others.

Some jobs will change or become obsolete as data apps improve in the not-too-distant future. Administrative roles where people perform routine tasks may go the way of the dodo bird. But we are not there yet.

We do not yet have purpose-built platforms for developing high-performance data applications. modern data stack companies like Snowflake and other cloud vendors continue optimizing their platforms to create and run data applications. I think we will see some exciting solutions emerge from the approach. However, data apps built using the modern data stack are jerry-rigged affairs. They are complicated to build and require resource-intensive and time-sapping techniques. Also, we cannot produce data applications using SQL alone.

Data scientists, those high priests of modern computing, do some of their work using SQL and then tap Python or other programming languages to perform predictive analytics. They write algorithms that pull data out of relational databases and into huge feature tables for analysis using machine learning techniques. Once complete, probabilistic assessments lead to automated actions.

Today's approach works but is far from optimal. We could optimize business processes to minimize the load on computing resources if we could avoid pulling information from databases and into feature tables.

The idea is easier to understand when we compare the process of data application creation to how hybrid cars work. Pairing a SQL database with Python/predictive analytics is like the way auto manufacturers put two different engines in a hybrid car, one electric and one gasoline powered. In today's data world, a lot of people drive hybrids. What we need is the data equivalent of a Tesla.

My friend Molham Aref, his team at RelationalAI, and more than twenty top database researchers from universities worldwide worked on this problem over the last decade. They invented the *relational knowledge graph,* a new type of database that allows organizations to fully model business processes within the database. In essence, your data application *is* a knowledge graph model.

The way Molham sees things, a relational knowledge graph is an important addition to the modern data stack. His metaphor: it is like accelerating the performance of your computer by adding a graphics processing unit (GPU) alongside your conventional central processing unit (CPU)—which is the set of chips that execute instructions.

Molham and his colleagues published hundreds of papers that examine the characteristics of relational mathematics. They also developed a new generation of relational algorithms that help support relational knowledge graphs. These improvements make this new generation of database possible.

At the highest level, a knowledge graph is a database that models concepts, their relationships, and the associated rules and constraints. It is a network of interlinked descriptions of objects, events, situations, and ideas. The heart of the knowledge graph is a model where the descriptions of entities are something that both people and computers can understand.

Relational knowledge graphs utilize an element that other knowledge graphs do not possess. They rely on relational mathematics to build data models, allowing business analysts and data scientists to describe the details and workflow of their business and encode those concepts directly into the database. This produces a relational knowledge graph model, which explicitly describes all the constraints and rules required to run the process. The model is executable and takes the place of the program typically written in a language like Java or Python.

In a SQL database, objects called tables combine a lookup key and multiple related values. Essentially, the key and values (columns) merge into a table to make it easier for the database to work with the data. With SQL, an association exists among a customer ID that includes a customer name, address, city, state, email, and phone number within a single table.

In contrast, a relational knowledge graph makes it possible to model all business process details. Unlike the tables used by SQL databases, a relational knowledge graph stores data in a format called *graph normal form*, which keeps each key-value pair in a separate object.

The process makes it possible to store a customer ID and that person's name separately from a customer ID and an email address. This ability offers more flexible data organization, but it also means many separate database objects must be combined to run a query. SQL databases cannot handle all these items because they use an older, less-efficient generation of relational algorithms. Storing the data in *graph normal form* makes it possible for relational knowledge graphs to define the business model in detail.

I see a colossal technology pivot coming in the years ahead. The modern data stack is powerful but immature and insufficient to handle the data management and analytic tasks we want to accomplish in the future. So, I see the ecosystem evolving to combine a maturing modern data stack with relational knowledge graphs and machine learning foundation models.

Just weeks after I left Snowflake in the spring of 2019, I received an unexpected call from Molham. I had not heard of him or his company,

RelationalAI, which was new and working on top-secret technology. Molham read about my departure from Snowflake and wanted to bring me into his orbit. He invited me to his house in Woodside, California, sat me down in front of a whiteboard, and explained his goal. It was a captivating four-hour meeting, but I understood maybe 20 percent of what he shared. Still, I understood enough to get excited about what he was doing.

After my first meeting with Molham, I had a couple more appointments with him and his technical team. About a year later, I engaged fully with them. I have weekly meetings with Molham and his team, joined their board of directors, and invested in their Series A and B rounds of venture capital funding. I am helping them build the first version of their relational knowledge graph technology and bring the product to market.

Molham became a technology rebel at an early age. He was born in Lebanon, but his family left the country after civil war broke out in the mid-1970s. His father's employer, DuPont, moved the family to Geneva, where, at age 13, Molham learned to write programs in Basic on his school's DEC VAX computer. He fell in love with computers, especially the Apple Macintosh, but he didn't see himself becoming a software programmer. That seemed boring to him. Instead, he sought to make computers more accessible to him and everybody else. He felt the world should not need armies of software programmers giving computers tiny step-by-step instructions. His dream is to tell a computer *what* to do without telling it *how* to do that.

The family later moved to the United States, where he got undergraduate and master's degrees in computer engineering and electrical engineering at Georgia Tech. He climbed the leadership ladder in several companies before launching several of his own. His last major stop before RelationalAI was as founder and CEO of LogicBlox, where he developed a database that used *declarative programming* to work with data.

With declarative programming, instead of requiring programmers to describe how to perform a task step-by-step with procedural code, the database figures out the optimal way to get the job done using the relational mathematics and algorithms built into the system's semantic optimizer. In other words, a business analyst or data scientist declares the *what,* and the semantic optimizer figures out the *how.*

Although LogicBlox had some limitations, it was Molham's first attempt to build a relational knowledge graph system. It allowed him to understand better how to embed business logic in a database. Molham sold LogicBlox to Infor in 2016. After selling the company, he kept working on declarative programming and relational knowledge graphs. And out of that work came the underlying concepts for building relational knowledge graphs.

Tech giants developed the most common knowledge graphs that help organize, search for, analyze, and present information found on websites. A good example is how Google automatically pulls critical pieces of data from the highest-quality search results and presents them in summary form in a sidebar.

The illustration below provides an example of the Google Knowledge Graph, providing useful details about the Museum of Modern Art in New York.

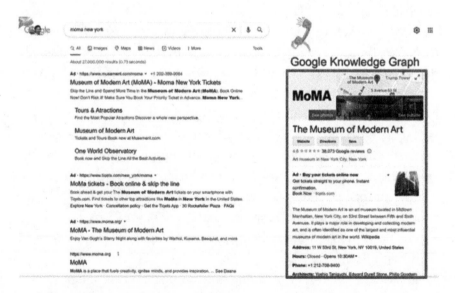

This knowledge graph can identify the most critical information on a web page, categorize it, and extract it. From there, the knowledge graph can summarize the information elsewhere on the web page or another page, so people easily understand it.

Today's knowledge graphs are bespoke systems purpose-built by the world's largest tech companies. There is not yet general agreement in the industry on how to build the next generation of knowledge graphs. Figuring this out is Molham's lifework, and he believes that relational mathematics is the solution to building a general-purpose knowledge graph system.

Because a relational knowledge graph includes the database schema and the business logic, it fully executes a business model without requiring procedural code running in an application server. The approach is much simpler than the typical hybrid approach of today's data applications.

A fundamental problem with today's approach is that business logic gets scattered into multiple systems. The same calculation is often duplicated and implemented numerous times in different applications and within BI dashboards. Putting all the business logic in one place, the knowledge graph model prevents errors and makes the code much easier to maintain.

Molham believes—and I agree with him—that applications built this way are more efficient with peoples' time, require fewer computing resources, increase profit margins, and accelerate growth for organizations that use them. Over time, he believes these systems will be so powerful and easy to use that company leaders can model their businesses for themselves. Less-technical staff can describe what they want to achieve in simple terms and let the system figure out how best to meet those goals. As a result, businesspeople and data professionals can use the knowledge graph to automate a lot of the drudgery that exists in their jobs today. It will allow them to focus their precious time on the most crucial business problems.

For passionate technologists who want to know how the technology works, here's a little more detail: RelationalAI's engineers are using the Julia programming language to build their relational knowledge graph system so they can dramatically reduce the manual coding required. They developed Rel, a declarative language used within the knowledge graph to enable the development of new data apps. Unlike traditional languages that necessitate orderly line-by-line coding, programmers using Rel do not need to specify the order of operation. Instead, Rel statements are directly associated with the business model and the semantic organizer within the knowledge graph system decides the optimal order automatically.

Think of it this way: When organizations create business applications using conventional techniques, they implicitly perform business modeling. The models for how businesses work exist in the brains of business leaders and analysts within organizations. Programmers must convert those mental models into software code that controls how we organize and manage business processes. The effort requires the combined expertise of someone who understands the business and a programmer who understands the code. Problems ensue if the programmer does not correctly translate the company-specific knowledge and intention into the code.

In contrast, the relational knowledge graph system helps business leaders, analysts, and programmers perform business modeling using a more formal and manageable process. We are moving from *implicit* to *explicit*.

With explicit modeling, the businessperson describes how they want the company to run and the programmer translates those intentions into code, but the big difference is that the businessperson can see the rules defined by the code, and it won't look like hieroglyphics. They will understand the code well enough to know if the translation from ideas to code is accurate and describes the constraints correctly. This way, staff can discover mistakes in implementation before deploying a new data application in the organization.

How does this apply in the real world? When I talk to technology leaders at enterprises, time after time, I hear that governance remains a fundamental problem for cloud data analytics. A colossal governance headache can occur when the public cloud combines data from multiple sources, provides access to it, and shares it with business partners.

Companies must have specific mechanisms in place to manage data governance effectively, such as the ability to inventory all data assets, the ability to model data and business concepts, and the ability to handle issues in data quality. Most importantly, they must also provide secure access control and rights management.

Access control over data is critical for business compliance and risk management. Currently, these capabilities are provided in a highly fragmented way by many different vendors. These products do not work particularly well with each other. There is no underlying database platform that fulfills all data governance needs. SQL is insufficient for governance applications because it cannot support the complex relationships that governance requires. So, using the modern data stack's SQL database to manage data governance properly is not currently possible. No wonder customers struggle with this challenge!

Relational knowledge graphs provide a solution because they support a semantic layer that encapsulates all the necessary capabilities required for governance. Relational knowledge graphs can play this role because they can model company concepts, their relationships, and the associated business rules and constraints. By starting with a model of the business, relational knowledge graphs provide an architected approach to managing security, structure, and data quality.

Knowledge graphs can be very general purpose. They can also be limited to a data model or contain the complete structure of an entire process. They can model a product, a business, laws, government policies, and business contracts.

When relational knowledge graphs become common, models will be defined explicitly and continuously updated with new information that reflects actual behavior in the real world. Applications built on them will provide predictions about future results and take action based on data that the knowledge graph model informs. Knowledge graphs will also support foundation models for training purposes and as reference sources.

This is exciting stuff. Technology must evolve rapidly to gain the full benefit of the data we collect these days. I believe that declarative relational modeling is one of the vital approaches. In the history of computing, relational approaches emerge again and again, challenge conventional wisdom, and win. I think the relational model will win once again.

In the future, I expect other companies will duplicate the work Molham, and all the engineers and scientists helping him, put into RelationalAI. The newly published relational algorithms and techniques that underlie the relational knowledge graph allow others to re-create the technology. This approach to publishing fundamental research is the same path that has allowed multiple SQL database vendors to compete over the decades.

SQL databases have remained in use for over forty years. If we compare relational knowledge graphs to the maturity of SQL, we are still in the very early days. SQL's success and longevity demonstrate the relational model's importance and enduring value. SQL served us well, but the time has come to augment it with a new approach. While I am confident that SQL will help us for decades to come, the days of SQL's dominance are numbered.

CHAPTER 11

THE FUTURE OF SMART MACHINES

The pursuit of making machines intelligent moved along slowly and sporadically through the early part of my career. Over the past decade, though, the pace advanced to a gallop. I distinctly remember the moment I realized that AI could do work that previously required the minds of brilliant people.

In early 2022, I met with Scott Guthrie, executive vice president of Microsoft's Cloud + AI Group, which includes Microsoft Azure, SQL Server, Power BI, and the GitHub team. While I am no longer a Microsoft employee, I have maintained a consulting role with Scott since 2021. The meeting was a review with leaders of GitHub, which Microsoft purchased in 2018. The GitHub platform hosts open source software development projects, and as of late 2022, it reported serving more than 94 million developers and more than 200 million repositories.

One of the platform's new features is Copilot, an AI technology that uses a machine learning model to assist developers with coding and functions in real time. It is available on both GitHub and Microsoft's Visual Studio developer platform.

At the beginning of the meeting, we reviewed the written report that helped focus our discussion. One bit of information jumped out: Copilot software wrote 40 percent of the code GitHub developers checked in, demonstrating AI's remarkable potential. When the discussion started, I pointed out that few technologies in the history of computing have had that kind of impact on people.

Copilot is an example of a machine learning approach causing excitement throughout techdom: *foundation models*.

Researchers at Stanford University's Institute for Human-Centered Artificial Intelligence coined the term. In 2021, they published a 214-page report, "On the Opportunities and Risks of Foundation Models,"[7] defining foundation models and laying out a strategy for making the most of them.

Their definition is disarmingly simple. Foundation models are computer models trained on large amounts of data and adaptable for many applications. Language is the underpinning of all foundation models, including the ability to interact with people using everyday English. Foundation models typically derive from deep learning and transfer learning, two established machine learning techniques. Stanford researchers say that "their scale results in new emergent capabilities, and their effectiveness across many tasks incentivizes homogenization." Homogenization, in this case, means that various methods combine to build a wide range of applications. The emergent capabilities of foundation models are poorly understood, so the folks at Stanford say the scientific community and the tech industry should proceed with caution before deploying foundation models broadly throughout business and society.

---

7   Bommasani et al. "On the Opportunities and Risks of Foundation Models." Stanford Institute for Human-Centered Artificial Intelligence, 2021. https://crfm.stanford.edu/assets/report.pdf.

The best-known foundation models now are BERT, DALL-E 2, Stable Diffusion, GPT-3, and its ChatGPT application. BERT is a technology produced by Google researchers that shows amazing results for many natural language processing tasks. DALL-E 2 and Stable Diffusion are AI programs that convert text descriptions into images. The OpenAI technology laboratory in San Francisco named DALL-E 2 as a takeoff on the famous surrealist artist Salvador Dalí. OpenAI also created Generative Pretrained Transformer version 3 (GPT-3). GPT-3 uses deep learning and the massive amount of information on the internet to produce humanlike text sentences and paragraphs. The chatbot ChatGPT, based on GPT-3, burst into the mainstream in late 2022. It helps in tasks like writing software code, drafting document summaries, and answering questions on any subject.

OpenAI is a fantastic organization. It began in 2015 in San Francisco as a nonprofit research lab. It had a wonderful group of high-powered backers, including Elon Musk, Reid Hoffman, Peter Thiel, and others, with additional funding from other businesses like Amazon and Microsoft. The original idea was to invest billions of dollars to create machine learning models that would be so vast and powerful that they would operate like public electricity utilities. Later, the organization restructured to create a nonprofit foundation and a commercial arm, which it uses to market and license its technologies. One of the critical things to understand about OpenAI is that its overarching goal is promoting and developing AI in a way that benefits humanity.

The CEO of OpenAI is Sam Altman, little known outside Silicon Valley before late 2022. But with the explosion of attention around GPT, ChatGPT, and DALL-E 2, he emerged as one of the leading tech visionaries today.

I met Sam late in my Microsoft stint while running the Server and Tools business. I presented to a group of Silicon Valley venture capitalists at Microsoft's Valley headquarters, and my team arranged to have Sam Altman join me on stage to run a demo. At the time, Sam was CEO of a tiny startup, Loopt, based in Mountain View, which provided a service enabling smartphone users to share their location with others. Loopt was an early adopter of Microsoft's 64-bit server products.

Sam, in his mid-twenties then, did a great job demoing Loopt on our platform. He impressed me. However, I had no clue that he would later become one of the leading technology and thought leaders in what is shaping up to be the era of artificial intelligence.

In 2011, Sam became a part-time partner in Y Combinator, the famous Silicon Valley startup investor and incubator. When Loopt sold in 2012, he focused on Y Combinator and was named president in 2014. He continued in various management and investor roles there even after he started as the first cochair and later CEO of OpenAI.

Sam and some of the OpenAI cofounders laid out their goals in a blog post[8] in 2016. They mainly focused on basic research into machine learning and creating a robot capable of doing housework, intelligent agents that could play computer games, and a chatbot capable of answering questions and conversing with humans. These would be the test beds for some of their more significant ideas.

Since then, OpenAI has shared a steady stream of platform technologies. Many are available to developers through application programming interfaces (APIs) or released as open-source software projects. OpenAI Gym, a platform for reinforced learning, and Universe, a tool for measuring an AI system's general intelligence, came in 2016. The company introduced GPT-3 in 2020, DALL-E in 2021, ChatGPT in late 2022, and GPT-4 in March 2023.

Sam was a guest on innumerable panels and fireside chats captured on video and streamed online. You can get a feel for his thoughts if you watch some of them on YouTube, but here are some crucial takeaways.

After the establishment of OpenAI, Sam explains, the conventional wisdom in AI circles was how machine learning improvements would

8   Stutskever, Brockman, Altman, Musk. "OpenAI Technical Goals." OpenAI.com, June 20, 2016. https://openai.com/blog/openai-technical-goals/.

first center on teaching models to do specific things using a particular set of data. Over time, they would learn more and more about human knowledge. It turned out the opposite was true. Foundation models train on enormous volumes of data, information, and knowledge on the web. Fundamentally, he says, the huge leap came because we developed models that can truly learn. They rapidly improve when we throw more computing resources and data at them.

He believes we have most of the technology techniques we need to reach artificial general intelligence (AGI), the next major milestone in computer intelligence. Sam defines AGI as a computing system with intelligence equivalent to a median human. Building an AGI is a crucial focus for OpenAI. Sam thinks that future models that OpenAI and others develop will improve dramatically. He says we need to use more computing resources and data to train the models to get there.

The techniques used by foundation models like GPT are surprisingly straightforward. These "large language models" are trained by blocking out words within a sentence and asking the neural network to predict the missing word. When it guesses correctly, that strengthens the weight of those connections within the neural network. By training the model against accessible sentences on the internet, it learns to write. The program is so good that people cannot tell if a human or a machine wrote its output. That milestone means it can pass the so-called Turing Test for demonstrating machine intelligence, defined in 1950 by English mathematician and digital computing founder Alan Turing. GPT-3 trained with the help of Azure infrastructure. Microsoft has licensed exclusive use of GPT-3's underlying model, but others can use the public API within their programs.

Microsoft also has internal groups working on foundation models, which are at the core of Microsoft's AI strategy. In the past couple of years, these technologies have allowed Microsoft to produce some of the best results among its industry peers in improving computer speech, language, and image processing. My friend Xuedong Huang (XD, for short) is a Microsoft technical fellow and chief technology officer for Azure AI. XD has focused on improving Microsoft's AI capabilities throughout his long career. He calls the company's strategy "integrative AI."[9] Through integrative AI, Microsoft integrates all kinds of data and relevant AI techniques, including foundation models that address the challenges of handling speech, text, and images. "The era of foundation models is here," says XD.[10]

XD says his team draws inspiration from Johannes Gutenberg, the German inventor of the printing press. Gutenberg's invention resulted in humans sharing knowledge more broadly than ever before. By breaking down words into letters using movable type, Gutenberg could easily mix and match them to replicate any book, proclamation, or pamphlet—eventually giving rise to newspapers, magazines, and printed materials. Similarly, Microsoft's code breaks down AI capabilities into smaller building blocks that combine to make computing systems more intelligent. For example, training a model to transcribe speech in multiple human languages improves the fidelity of English transcription.

Likewise, Microsoft's foundation models are built and used like the layers of a foundation underlying a physical building. The company's scientists have created a collection of foundation models for three sense-making modalities—monolingual text, audio or visual sensory signals, and multilingual text. That is the first layer of building blocks. They then mix and match the models to create larger foundation models within each modality to improve results for a single data type. That is building block layer #2. Then, for applications that benefit from a combination of modalities, they plan on bringing all their foundation models to bear to achieve cross-domain learning, spanning modalities and languages. "The intersection came from the realization that one dimension was not sufficient," says XD. "Just like with people, it's important to teach machines using all of the senses."[11]

---

9   Huang, Xuedong. "A Holistic Representation Toward Integrative AI." Microsoft. com, October 20, 2020. https://www.microsoft.com/en-us/research/blog/a -holistic-representation-toward-integrative-ai/.

10 Xuedong Huang, in conversation with Bob Muglia and Steve Hamm. August 31, 2022.

11 Ibid.

XD took a fantastic journey from the hinterlands of China to the peaks of AI innovation. He grew up in a remote part of China's Hunan Province. Chairman Mao Zedong's Cultural Revolution essentially wiped out university education, but XD benefited from the liberalization brought by Mao's successor, Deng Xiaoping, who restarted the universities. Anxious to get going, XD dropped out of high school at age fifteen, took the national university entry exam, passed it, and enrolled in Hunan University. That launched him on an academic path that took him to Tsinghua University in Beijing for his master's degree, Edinburgh University in Scotland for his PhD, and then to Carnegie Mellon as a postdoc and junior faculty member.

Starting in undergraduate school, he focused on spoken language processing. Computer keyboards were a challenge for Chinese-speaking people, and he wanted to help his country's people communicate with computers by talking to them. For decades, this was XD's specialty. Today, Microsoft's most popular productivity applications use high-quality live speech captioning. For instance, a Microsoft Teams video call provides real-time, accurate transcription. Because speech has so many variables, natural language processing (NLP) pioneers trained their models with massive amounts of samples, so "the foundation model idea goes back to speech," XD says.[12]

I remember a video that Microsoft's marketing team and I made for our TechEd conference in 2007. At the time, NLP accuracy was abysmal, and it was always highly questionable whether an NLP demo would work on stage. Our video was a takeoff on the popular 1980s movie *Back to the Future*, in which where a couple of ill-matched buddies traveled back and forth in time in a DeLorean sports car. In the Microsoft video,[13] I am booed offstage by an audience of developers tired of the company's never-ending vision-of-the-future presentations. I dressed like actor Michael J. Fox's character. Then Doc—the

---

12 Ibid.

13 Whsfritz, "TechEd 2007 Keynote opening (Back to the Future style)." Youtube, September 4, 2010. https://www.youtube.com/watch?v=KxMrBuluEJ8.

actor Christopher Lloyd from the film—took me on a trip into the past to view some of Microsoft's off-target data-management schemes, including Integrated Storage and Cairo. Then we time travel into the fictional future to see how our earlier visions played out. At one point, Lloyd speaks to the computer in the DeLorean, asking it to take us to 2015. A robotic voice responds: "Did you say, 'Happy Halloween?'"

It was easy to make fun of NLP in those days. Not so today.

When XD started in the AI field, the cutting-edge natural language processing technology was "expert" systems that used rules and reasoning to predict what word was spoken. Shortly after, computer scientists relied on statistical methods for calculating probabilities and converting speech to text. Then came deep learning, which accurately converts speech inputs to text outputs. In 2016, using these techniques, XD and his colleagues demonstrated a speech-recognition system that had achieved parity with humans. These techniques can improve language translation quality, image identification, and analysis.

That brings us back to foundation models. Computer scientists realized that by feeding machine learning systems massive amounts of sample data, then using deep learning to draw inferences and make predictions, they could produce tremendous advances in system performance. Then they combined data types and used the *transfer learning* technique to create links between knowledge gained using different modalities, thereby creating huge models with many uses. Multimodality foundation model systems are still in the toddler stage. Still, in 2017 Microsoft built a single foundation model for speech translation that now addresses more than 100 languages and works with high-fidelity audio and lower-quality cell phone messages and conversations. It packages the technology into an API to make it easier for developers to include these capabilities in their applications. Microsoft has been using foundation models since 2017, but before Stanford researchers coined the term, Microsoft called them unified models.

Deep learning, an essential capability of foundation models, has emerged over the past decade as the most effective way to train computing systems to mimic human thinking. The idea is that machines can best process complex datasets when the work takes place in layers. Each layer extracts a different kind of information and compares it to training samples or previous learnings to predict the meaning of the new data. For instance, the lower layers may identify simple curves and shapes when processing the images in a security video while the higher layers identify street names, license plate numbers, and even the faces of individuals. These processes are like a series of sieves that remove ever-finer material from dirt.

The term "deep" refers to the fact that these algorithms might employ fifty processing layers or more. If you want more detail, Yoshua Bengio, a professor at the University of Montreal, laid out the basic concepts underlying deep learning in his 2009 paper, "Deep Learning Architectures for AI."[14]

Yoshua is one of the so-called godfathers of deep learning, along with Yann LeCun of Meta and New York University and Geoffrey Hinton of Google and the University of Toronto. The trio teamed up in 2015 to publish a seminal research paper, "Deep Learning."[15] Three years later, they earned the A. M. Turing Award, which is like the Nobel Prize for computer science. The recognition resulted from their development of the mathematical and computer science under-pinnings for deep learning and for improving the effectiveness of arti-ficial neural networks, often combined with deep learning algorithms.

Foundation models use artificial neural networks to process the bits and bytes representing their operational data. These so-called "neural nets" are loosely modeled on the human brain. Each consists of mil-lions or billions of individual nodes, which function like neurons in the brain to share information. The connections between the network nodes are like the synapses in the brain, transmitting signals between nodes. Neural nets and deep learning architectures organize in layers. Nodes on one layer connect with each other and to the layers above

---

14 LeCun, Y., Bengio, Y. & Hinton, G. "Deep learning." *Nature* 521, 436–444 (2015). https://doi.org/10.1038/nature14539.

15 Ibid.

and below it. In an artificial neural network, data processing occurs on powerful integrated circuits—either Graphics Processing Units (GPUs) like those developed initially for the computer game industry or specific integrated circuits built for AI applications.

A vital feature of these foundation models is the single, end-to-end approach for training across all model layers, simultaneously enabling training for the entire multilayered model. Interestingly, automatic differentiation and the calculus derivatives I described in Chapter 9 are used to adjust the weights of the connections across the neural networks during training.

When Yoshua wrote his 2009 paper about deep learning, we did not yet have learning algorithms to understand real-world scenarios and describe them in natural language. And we did not have algorithms that could discover the visual and semantic concepts necessary to identify most images on the web. Thanks to deep learning algorithms and neural nets, now we do.

Today, many computer and data scientists are operating on the cutting edge of emerging technologies and possibilities. Frankly, I am not always clear on the advantages of one novel AI approach over another. But I have plenty of company. Even the experts do not always agree on what techniques will produce the next significant performance breakthroughs in AI.

Recently, I asked Molham Aref of RelationalAI to jump on a call to discuss the future of data analytics technologies.[16] He took me on a historical tour of AI. We then discussed the challenges computer scientists face in making machines even more intelligent than they are today. Together, we developed this breakdown of six key attributes of machine intelligence:

- **Sense:** The ability of machines to recognize and process all types of data, including structured, semi-structured,

---

16 Molham Aref, in conversation with Bob Muglia and Steve Hamm. October 11, 2022.

text, images, video, and sound. Machine learning can help to unlock and identify the contents of all data types. While there is much work ahead, machines can excel at sensing.

- **Learn:** The ability to use algorithms and statistical methods to draw inferences from data without following explicit instructions. While global-scale foundation models have a long way to go, their ability to navigate vast amounts of human-generated content helps them demonstrate knowledge about our world that can dramatically exceed human capabilities.

- **Reason:** To draw conclusions from diverse collections of data based on data and background knowledge. Traditionally, software programmers code most logic. In the future, machines will generate the logic required to reason based on simple instructions from humans. Today's foundation models are brilliant at pattern recognition and auto-completion. However, they don't understand how the world works because they don't yet have a model for the world. As a result, they can produce wrong or meaningless results. Improving machines' ability to reason is a primary goal for computer scientists today.

- **Plan:** To optimize a process for accomplishing a goal. While today's systems can perform relatively straightforward tasks like planning and optimizing a route through traffic, more complex planning tasks are still out of reach. They require an understanding of context that is lacking. As models develop a greater understanding of the physical world around us, their ability to reason and plan will improve.

- **Adapt:** To gather and process data on a continuous basis and adjust automatically to changing conditions. While people are capable of continuously learning, today's models must be trained using an offline, compute-heavy batch process. This means that models must go through expensive retraining processes to stay up to date. The infrastructure used to build models must evolve to support continuous and incremental learning.

- **Act:** To take action based entirely on the model and changes
  in data without explicit instructions from people. This is the
  future of intelligent data applications and robotics. Every autonomous robot or agent contains a "behavior model" that makes
  decisions as the system performs its tasks. Today, these behavior
  models are relatively unsophisticated. They must improve considerably for life-critical tasks such as autonomous driving.

Computer scientists need to make progress in all of these areas to
deliver on their long-held dream of creating artificial general intelligence—the ability of machines to think as well as or better than
humans.

I do not know precisely how we will build the next generation
of intelligent machines. By nature and experience, I am a program
manager, not an engineer or architect. I look at what people are doing
worldwide to identify what needs to be added to computer systems to
help people achieve their goals more effectively or efficiently. Consider
the attributes I laid out above to be a specification for the future of
smart machines. I leave it to computer scientists, math geniuses, and
future datapreneurs to finish the job—and I am confident they will
succeed. Even though there is a lot of uncertainty about what comes
next in AI, the computer science community is anything but bogged
down. There is a massive amount of inventing going on.

At Microsoft, for instance, XD is open to experimenting with various techniques. "We want to be open-minded," XD says. Incorporating
foundation models technology from OpenAI and others across the
Microsoft product line will be a big step. Other companies are taking
similar approaches, and we can expect Google to follow.

XD believes that improving the interfaces between humans and
machines is critical in making machines more intelligent. He foresees
a paradigm shift coming as big as the change from the command-line
interfaces of early PCs to the graphical user interfaces produced for
the Apple Macintosh and Windows in the 1980s. ChatGPT is an early
example of this, and while it is still unclear how this paradigm shift

will evolve, XD believes this new generation of computer-to-human interfaces will take the form of avatars that look, think, and sound a lot like you and me.

Meanwhile, the Godfathers of Deep Learning are working hard on next-generation models. In 2022, Yann LeCun published a paper, "A Path Towards Autonomous Machine Intelligence,"[17] in which he wrote about the challenges ahead for computer science and proposed a new approach to building foundation models that would enable them to reason, predict, and plan across multiple time horizons. Today, foundation models are not good at reasoning or forecasting the future. They comb their vast portfolios of information to gather clues about a situation or a problem to solve, but they do not truly understand what they "see."

In his recent paper, Yann posits that the core capability that AI lacks is common sense. Because we humans possess common sense, we can learn complex skills quickly, predict the consequences of our actions, and respond rapidly to situations we have never experienced before. Not so for machines. He proposes building a new generation of intelligent agents powered by "world models"—vast collections of models from different domains that tell an agent "what is likely, what is plausible, and what is impossible." Yann proposes that in addition to vast portfolios of information and deep learning techniques, these intelligent agents need what he calls a "cognitive architecture." The cognitive architecture consists of modules, including a perception model, a predictor, a trainable critic, and other capabilities that run parallel to the biological features that provide people with common sense.

Yann does not lay out a detailed path forward. Instead, he presents his future vision, raises thorny questions, and suggests novel solutions. He is a bit like baseball slugger Babe Ruth, who famously pointed to the center field fence at Wrigley Field during the 1932 World Series,

---

17 Yann LeCun, "A Path Towards Autonomous Machine Intelligence." June 27, 2022. https://openreview.net/pdf?id=BZ5a1r-kVsf.

indicating that he planned on hitting the next pitch there—and then did so. Will Yann and the computer science community hit a home run that propels AI forward? I would not bet against them.

We see significant progress in foundation model reasoning. Early foundation models did not have a sense of time. For instance, they could not correctly answer the question: "Did George Washington own an iPhone?" ChatGPT correctly answers that question and will tell you why not. These improvements required a combination of unsupervised learning from earlier foundation models and reinforcement learning from human feedback, including annotated training to further refine the model.

What should we expect from foundation models in the coming years? The industry is rapidly evolving new generations of foundation models. GPT-4, released during the editing process for this book, provides another leap forward. Yet, as is typical with new technologies, gaps remain. One challenging gap is the data types used by today's foundation models. Text and speech were early areas of success for foundation models, and, more recently, computer scientists have demonstrated rapid progress with images and video. However, it is early days for foundation models that work with structured and semi-structured data. To master these data types, foundation models must evolve to handle the graph-oriented relationships inherent in this data.

Deep learning models designed to work with structured and semi-structured data are called *graph neural networks*. Like a relational knowledge graph, the data organizes as sets of related keys and values. Because graph neural networks and relational knowledge graphs build from relations, the new algorithms developed for relational knowledge graphs can apply to graph neural networks.

Another major weakness of current foundation models is that they only train on publicly available information. Data behind a corporate firewall is proprietary and thus inaccessible to these foundation models. We need a new generation of infrastructure to support companies that want to customize models for their businesses.

It is too soon to tell how this will unfold, but researchers around the globe work hard on these problems, and their progress is quite astounding.

Before we leave the topic of machine intelligence and this chapter, I want to explore another aspect of animal intelligence, sentience, and raise a provocative question: will machines ever become sentient? The meaning of the term is a bit muddy. Sentience is the ability to perceive things through physical senses, combined with the ability to experience and read emotions. Self-awareness is also frequently included in the definition.

Computers already possess analogues of human physical senses. I included that capability, sensing, as the first of the attributes of machine intelligence. But they do not yet experience emotions and are not fully self-aware.

For fun, I asked Molham what it takes for machines to achieve self-awareness. Molham answered, "There are two hard questions in the universe. The first is how it all started; what was there before the Big Bang? Physicists and astronomers are still struggling with that one. The other is sentience; at what point do animals and machines become self-aware? I'm not a philosopher, and I have no idea how we're going to answer either question."

I love dogs and believe they are self-aware. For example, our dog, Emma, is exquisitely attuned to her relationships with us. Her emotions are complex. She barks at the sound of a stranger but patiently waits when she hears my wife's car. She is sad when we leave her alone and vigorously greets us when we return. Anybody who owns and loves dogs knows that they have emotions. But does Emma realize she is a dog and we are people? I would say yes, but probably not with the clarity commonly understood by people.

It appears that sentience among animals exists on a spectrum. All animals, even a cockroach, can sense and perceive on one end of the spectrum. In the middle, some animals can read and feel the emotions of other living things. Recent scientific studies have shown that many animals, including dogs and dolphins, have well-developed emotional intelligence. If you want to learn more about this, I suggest reading a fantastic article published in 2022 in *National Geographic,* "What Are Animals Thinking?"[18] On the far end of the spectrum, we humans possess the capacity to experience many sublime emotions, including spiritual feelings. The bottom line is that scientists are concluding that the inner processes of animals are very much like those of humans. The main difference is we can express our feelings in language, and they can't.

Will machines gradually become more sentient on a similar spectrum? Will they come to experience analogues of fear, elation, empathy, and grief—even love? These are fascinating questions, partly because they force us to think about what technological advances would push machines over the line from cold and calculating rational intelligence into emotional intelligence and self-awareness. The traditional answer is that they can only develop these capabilities if we program them.

But wait a minute! Today's foundation models and deep learning systems possess emergent qualities. These models learn things and spot patterns without being directed by us. One experiment[19] run by Rakesh Chada, an applied scientist at Amazon-Alexa AI, demonstrated the emergence of neurons that accurately express complex thinking. The neural network used in this experiment was simple. It contained only 4,096 neurons, but the model used 82 million Amazon product reviews for training. Amazingly, a neuron emerged from this simple model that captured the reviews' sentiment. Imagine the inferences made by foundation models with billions of neurons!

As we introduce progressively more intelligent models, will AGIs gain humanlike self-awareness and consciousness? It seems likely that we will soon find out.

---

18 Yudhijit Bhattacharjee, "What are Animals Thinking?" *National Geographic,* September 2022. https://www.nationalgeographic.com/magazine/article/what-are-animals-thinking-feature.

19 Rakesh Chada. "The unreasonable effectiveness of one neuron." Rakesh Chada's blog, 2017. https://rakeshchada.github.io/Sentiment-Neuron.html.

Foundation models are changing the way people work with computers. Computers can now communicate in English and other native languages. Future versions of these systems promise to progressively improve the performance and capabilities of these large-scale foundation models. As these systems gain greater intelligence and ultimately demonstrate AGI capabilities, they will create a new world of possibilities. With these possibilities come significant ethical challenges.

CHAPTER 12

A NEW TYPE OF SOCIAL CONTRACT

While growing up in suburban Michigan, I was the nerdy sort of kid who loved math and science—and, not surprisingly, became the AV guy who ran the projector at school. I loved tinkering with electronic gadgets. My other great love as a teenager was science fiction, particularly the work of Isaac Asimov. I read hundreds of his novels and short stories during my teens and twenties. It was not until I worked on this book that I began thinking more deeply about intelligent machines' ethical and moral implications. I looked back at what Asimov had written about the relationship between people and robots. In addition to featuring robots prominently in his novels and stories, he wrote dozens of essays about them. Azimov concluded they should be respected, not feared or controlled. He developed rules for how they should interact with humanity. Asimov was an optimist and a realist.

I am an optimist and a realist, too. To achieve a future where artificial intelligence benefits humanity, we need a new social contract that

governs the relationship between people and the emerging generation of AGI machines.

The concept of a social contract emerged in the Age of
Enlightenment, the period of rigorous scientific, political, and phil-
osophical questing that spanned the eighteenth century in Europe.
Traditionally, a social contract is an implicit or explicit deal between
the government and the people in a country where individuals surrender some of their freedoms and follow the rules laid out by governments in exchange for other benefits and maintenance of the social
order. Because social orders are under constant stress, the values, laws,
and regulations that embody social contracts require reexamination
and modification when new factors come into play.

The new social contract I have in mind would govern the relationship between people and smart machines—assuring that people are safe
and the newly emerging AGI entities align with our values and interests.
It would consist of a set of rules agreed to by the world's governments,
businesses, and other institutions defining what intelligent machines can
and cannot do and how people can and cannot use them.

When I look ahead, I see amazing things coming. We are at the
beginning of an intelligence revolution–similar to our world in the
1850s during the Industrial Revolution.

This computer science journey began in earnest in the 1940s and
1950s. It was not until the 1960s that universities began offering computer science degrees. Then, the PC, the web, the smartphone, big data,
the cloud, foundation models, and tremendous advances in artificial intelligence came in rapid succession. The science and industry that emerged
from the IT revolution are still changing rapidly but are also maturing.
Technology transforms our world, businesses, and our personal lives.

Now we are on the verge of another great lift. By harnessing AI and
other techniques to master the explosion of data we see today, we can
understand how the world works much more accurately and compre-
hensively. We can make better decisions and use Earth's resources more
responsibly. Throughout my career, I have focused on bringing technology to bear to help businesses succeed. As computer technology

developed, its potential impact for good or potential harm increased. I am hopeful these coming technological advances will improve the wellbeing of our species and the sustainability of life on this planet.

Over the past eighty years of the computing revolution, intelligent machines matched or bested one human capability after another. First, we created machines with expertise in a single domain. With the arrival of foundation models, the depth of AI systems' knowledge, the speed with which they react or predict, and the accuracy of their predictions are pretty darned impressive. They already demonstrate a recall of knowledge far beyond human capabilities. Now, we create machines with expertise in multiple domains. These large-scale machine learning models will dramatically lower the cost of intelligence, enabling new smarts and capabilities in applications and services of all types.

Remember the "arc of data innovation" graphic I shared in the book's introduction? Let's look at it again.

The Arc of Data Innovation

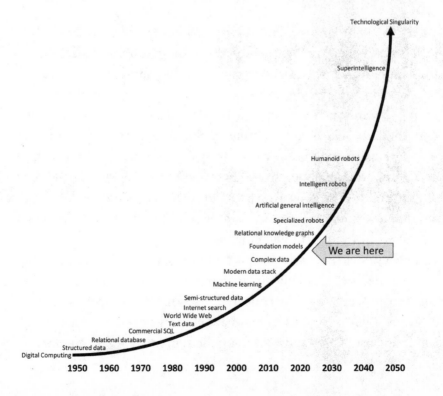

The earlier chapters of this book trace the arc of data innovation from the 1950s to the present. Each item along the continuum represents the period when a new concept or capability hit the mainstream. Look at where we are now, hovering over foundation models. I gave you my views on the potential for foundation models and relational knowledge graphs in Chapters 10 and 11. The next significant steps forward include artificial general intelligence (AGI), intelligent and humanoid robots, superintelligence, and technological singularity. Sam Altman from OpenAI uses the following definitions:

AGI: This will be achieved when a computing system possesses intelligence equivalent to that of a median human.

Artificial superintelligence: When a system possessing AGI is smarter than all of humanity put together.

Technological singularity: When AGIs working with people enables hundreds of years of progress in one year.

I see the 2030s and beyond as the golden era of robotics. Today's robots that make cars on assembly lines and clean up nuclear power plants after meltdowns are impressive. But the robots of the future will impress on another order of magnitude. They will be largely autonomous—because they cannot always take orders from computing systems located in the cloud. To make machines capable of autonomy, we will have to provide them with AI capabilities that are miniaturized and localized. Initially, these machines will serve single purposes like cleaning our floors, delivering packages, driving vehicles, and flying us around. Over time, more general-purpose robots that take on human characteristics and form will emerge.

I believe machines will possess artificial general intelligence within the next decade. It is only a matter of when. These prospects do not frighten me, but they do concern me. What is the societal impact in a world where smart machines are general-purpose, matching the capabilities of people and exceeding them in many ways? What ethics and rules will control these machines? In the future, it seems likely that robots will be capable of performing most physical tasks, and intelligent models within them will be capable of performing most intellectual tasks. What will people do if machines and AI systems do all that work?

As I said in the prologue, I am not an expert in AI, and I am not an ethicist. I am an engineer and a businessman. I do not have a clear answer to these issues. These questions will likely be among society's most critical policy issues in the decades ahead. Computer scientists, business leaders, government officials, academics, ethicists, and theologians must work together.

I believe people will develop solutions to the profound ethical issues raised by tomorrow's robots and intelligent machines, but I think the process will be messy. In history, every major technological advance has been used, for good and bad. Ultimately, though, common sense prevails, and society establishes laws and regulations that oversee the use of technology. This governance applies to everything from electricity to nuclear technology, and I believe the same will happen with intelligent machines.

These are hard questions but not new ones. Science fiction embraced them for decades. Indeed, there is no shortage of dystopian books and movies on the subject. Director James Cameron drove home the danger of a rogue, self-aware super intelligence in his 1984 film *The Terminator*. As you might recall, in the movie, an artificial neural network called Skynet is given control of the nuclear arsenal of the United States. One of the film's main characters, Kyle Reese, explains that Skynet assessed the situation and then "decided our fate in a microsecond: extermination," starting a nuclear war against humanity.

Other storytellers had more positive and balanced views of a world where people and artificial intelligence coexist.

The term "robot" was coined in 1920 by the Czech playwright Karel Čapek in his play *R.U.R.*, which stands for Rossum's Universal Robots. Čapek's play struggled with the moral and religious aspects of humans creating another intelligent being.

Isaac Asimov treated robots as machines built to help people. He gets credit for coining the term "robotics,"[20] defined as "technology that deals with the design, construction, operation, and application of robots." From his earliest writing, Asimov believed that robots would have "safety factors" programmed into them. In the 1942 short story "Runaround," which he published at the age of twenty-one, he begins his lifelong exploration of robotic ethics by introducing the Three Laws of Robotics:

> **First Law:** A robot may not injure a human being or, through inaction, allow a human being to come to harm.
> **Second Law:** A robot must obey the orders given it by human beings except where such orders would conflict with the First Law.
> **Third Law:** A robot must protect its own existence as long as such protection does not conflict with the First or Second Law.

He later added another law, which he called the Zeroth Law: A robot may not harm humanity, or, by inaction, allow humanity to come to harm.

These laws play a central role in many of Asimov's works. His short stories and novels explore the nuances of robots obeying the rules and the implication of this as they interact with far-less-perfect humans. The laws usually succeed in keeping robots in their place as servants to people. However, some stories describe how violating the laws can lead to problems.

Asimov was quite proud of his laws. In *Robot Visions*, one of his later collections of stories and essays, he wrote: "I have managed to convince myself that the Three Laws are both necessary and sufficient for human safety in regard to robots. It is my sincere belief that some day when advanced human-like robots are indeed built, something

---

20 Merriam-Webster.com Dictionary, s.v. "robotics" definition, https://www.merriam
-webster.com/dictionary/robotics. Accessed March 2023.

very like the Three Laws will be built into them. I would enjoy being a prophet in this respect."

While many people are familiar with Asimov's Three Laws of Robotics, he also created another set of laws that he felt should govern human conduct, which he laid out in an essay in *Robot Visions*.

Asimov's Laws of Humanics

First Law: A human being may not injure another human being, or, through inaction, allow a human being to come to harm.

Second Law: A human being must give orders to a robot that preserve robotic existence, unless such orders cause harm or discomfort to human beings.

Third Law: A human being must not harm a robot, or, through inaction, allow a robot to come to harm, unless such harm is needed to keep a human being from harm or to allow a vital order to be carried out.

Now, here we are, more than eighty years from when Asimov began laying out his prophetic ideas, and the discussion of rules for robots is no longer theoretical. Because of machine learning, computers have started making decisions for us, and truly intelligent machines are not that far off. We cannot put off dealing with the moral implications of machines making decisions that impact our lives. Asimov showed us the importance of deeply encoding ethics into the machines we create. In Asimov's writings, the laws are wired permanently into the "positronic brain" of his robots.

Today, machine intelligence manifests in ways that Asimov did not imagine. We use software, foundation models, and multiple deep learning layers rather than hard-wired artificial brains. Computing systems possessing AGI will run in cloud data centers, and robots will connect to them via the internet. In contrast to the hard-coded wiring of a positronic brain, computer scientists today must encode ethics and rules into fungible software models. The ability to easily change a software model to reflect different ethics starkly contrasts the world Asimov envisioned.

While incomplete, Asimov's Laws of Robotics and Laws of Humanics can serve as starting points for developing ethical and governance models to form the basis of a new social contract that governs machine behavior and human-with-machine interactions. People building future AGIs and robots will decide what values their product follows. That effort will require the work of many people and organizations across the public and private sectors.

There is another aspect of the social contract idea I want to touch on: the social safety net. This new era of intelligent machines will bring massive changes and disruptions affecting the economy, our institutions, and our lives as individuals. We must think deeply about the social consequences of these phenomenal advances in machine intelligence and begin planning responses to them.

One concern is inequity. How will humans earn a living when intelligent machines do most manufacturing, transportation, distribution, and knowledge-work jobs? In modern capitalism, tremendous gaps exist between the powerful and wealthy few and the not-so-powerful and not-wealthy many. These inequities could become even more pronounced as machines take over much of the work to drive the economy and improve human wellbeing. Corporations and their leaders will own these fantastic productivity tools, and most people will be customers.

Another concern is with access. Individuals with easy access to AGI systems will possess tremendous advantages in acquiring education, knowledge, financial resources, and other things that drive success. Unless access to these capabilities is available to everyone, the few will become even richer and more powerful. Modern societies could start to look like the feudal states of the Middle Ages. That kind of power-and-wealth dynamic is incompatible with democracy and unsustainable.

We can and will overcome these challenges, and the rising tide can lift all boats. But these issues will not solve themselves. We must think deeply about them and design solutions before the disruptions take full force.

In this book's introduction, I described the importance of values in running a business. As this book has developed, it has become clear that we must make values paramount in building and relating to smart machines.

For organizations to succeed over time, they must embrace a clear set of values. Mission statements formalize a company's purpose, and values guide the culture and employee actions. As we advance into the model-driven era, we must imbue computing systems with moral intelligence. Organizational and societal values must be a component of our computing systems. "Ethics models" need to define proper and inappropriate actions and provide boundaries for software applications and robots. Developing successful ethics models will be critical to the future relationship between people and machines.

Programmers manually code a subset of a company's values and business rules into many cloud-based applications today. These guidelines govern everything from product warranty policies to data governance. We did this quite intentionally at Snowflake. However, these policies and rules are far from comprehensive. As products evolve to become software services that deeply embed machine learning and foundation models, the actions taken by machine learning models will drive the behavior of those services. To control this behavior, we must encode ethics into service models. People need to monitor and update ethical controls in the near to medium term. In the long term, applications will adapt and evolve business rules without much human supervision.

Government policies will define some of these ethics models, and those policies will vary by country. We have seen Europe take a more aggressive role in regulating privacy, and I expect that to continue.

Other issues will be decided differently by different organizations. Issues of content moderation and censorship are front and center right now. Ultimately, society must establish policies for appropriate online human behavior and rules that govern the conduct of machines.

While the details will vary among technology service providers, clear rules and boundaries will emerge.

Autonomous robots, especially robots that drive on roads and carry people, must operate using clearly defined rules. The ethical challenges here are quite significant. What should a self-driving car do when faced with the choice of hitting a pedestrian or smashing into a concrete wall? This example is one of many scenarios where decisions that might result in harm must be specific and explainable. Accidents will happen, and courts will test them. Regulations will follow, and success requires clarity on the rationale behind decisions. Our current models and autonomous systems have not yet reached that level of sophistication, but future models must explain the ethics behind the decisions they make.

Foundation model-based applications such as ChatGPT are trained on the breadth of internet content, including all its sources and perspectives. But throwing such a wide net can cause problems. A lot of incorrect information lives on the internet, and there is no shortage of hate speech and propaganda created by malicious actors. How can we harvest valuable details while filtering or downplaying poor-quality and antisocial content? Fortunately, early examples of ethics models are emerging. GPT-3.5, used by ChatGPT, contains content moderation based on training and reinforcement learning from human feedback that helps to ensure that the questions asked and ChatGPT's answers are socially acceptable. People within OpenAI set the rules.

From my experience using ChatGPT, OpenAI is trying hard to incorporate commonsense ethical guardrails. But that approach to content moderation raises still more issues. OpenAI's judges of what is socially acceptable have their own biases, which can cause problems, too. Making the algorithms and training data used for content moderation and ethics models available for review could be a solution to this dilemma.

Foundation models are still a long way from becoming the all-knowing oracle. However, they show incredible promise in digesting human knowledge and providing valuable answers to complex questions. As foundation models develop, they must follow common sense and ethical values to fulfill their promise. It is incredibly early days for foundation models, but the pace of innovation is breathtaking. Watching this new technology develop in the months and years ahead will be fascinating.

In the introduction, I described an era of progress and prosperity. I think machine intelligence will introduce this new era, which could ultimately create a world of plenty with sufficient abundance to lift people out of poverty and reduce inequality. So, what is possible in a world of AGIs and superintelligence?

Intelligence, energy, and labor fuel our economy. AGI and artificial superintelligence promise to quickly reduce the cost of knowledge and intelligence.

Renewable technology can lower the cost of energy, and a new generation of smaller and safer nuclear power plants is under development. Recognizing that future AGI systems require power-hungry data centers, Sam Altman funded new types of power systems. Most interestingly, Sam invested $375M in fusion startup Helion Energy,[21] which promises to introduce commercial fusion systems within the decade. Fusion may be decades away, but short-term advancements are a fantastic development.

The third piece of the puzzle is labor, and the mechanical engineering of robots is further advanced than the intelligence needed to make them work. As artificial intelligence progresses, it will directly translate into smarter robots. These robots will eventually take humanoid form and help with various tasks. In time, reductions in the cost of labor, intelligence, and energy, opens many possibilities.

We will have two kinds of intelligent machines—smart robots and very smart AGIs.

Robots will be our servants. They will help us get stuff done. Think about care for the elderly. Today, many families must hire people at great expense to provide 24/7/365 care for elderly and infirm

---

21 Helion Energy: https://www.helionenergy.com/fusion-energy/.

relatives. In the not-too-distant future, robots will provide this service. They will cook, clean the house, move furniture, administer medication, help their human get to the bathroom, and provide companionship. A connection to AGIs via wireless network connections will give them access to the knowledge of a doctor, a nurse's bedside manner, and a psychiatrist's insight. Maybe they will do our taxes when we are sleeping.

One of my favorite TV characters from my youth was Rosie, the robot maid who minded George, Jane, and the rest of their futuristic family in *The Jetsons*, an animated science fiction sitcom on air starting in the early 1960s. I loved that Rosie kept the family in line and was so sassy. I think our future robots will have personalities, too, because we will enjoy interacting with them.

Unlike our robots, AGIs will not be servants. They will be as smart or smarter than we are, and we must show them respect. It is probably best to think of them as our peers! Because these machines will be so intelligent, they will help us understand the puzzles that confound us now—what came before the Big Bang, how to end poverty, how the mind and body work, and other profound questions.

Think about medical science, for example. Today, it takes a decade and billions of dollars of investment to bring a new miracle drug to market. With the assistance of AGIs, we will be able to diagnose unique illnesses in individuals and accurately and quickly develop individualized therapies—and we will not have to test drugs on animals or conduct clinical trials on humans.

Intelligent assistants will be indispensable companions, like smartphones are today. They will likely be embedded in smartphones and interact with us using natural language and voice. They will help us achieve what I see as humanity's purpose: to learn about ourselves and the world to make life better for ourselves, our families, and all humanity.

I end the book where I started—talking about the importance of clearly established values and ethics. The machines we create will someday exceed our capabilities, and they will embrace the values we define. They have the potential for profoundly positive benefits to humanity and help everyone lead happier and more productive lives. Let's build these machines with values based on a social contract—establishing a foundation for long-term human and machine collaboration.

EPILOGUE

In so many ways, Asimov was ahead of his time. He was a prophet. His robot stories and novels provided parables that describe the ethical and moral complexity of worlds where people and intelligent machines peacefully coexist. In his 1956 short story, "The Last Question," Asimov described his vision for the future of humanity. "The Last Question" was Asimov's favorite story, and it is mine, too.[1]

In the story, Asimov envisions generations of human-computer interactions stretching billions of years into the future. The vignette features progressively more intelligent computers: Multivac, then Microvac, then Galactic AC, then Universal AC, then Cosmic AC, then just AC—each designed and built by its predecessor. Humans ask the computers variations of the same question: "How can we reverse entropy?" In other words, how can we prevent the seemingly inevitable transition from order to disorder, where the universe's energy is exhausted? In each case, the answer is: "insufficient data for a meaningful answer."

---

1 "The Last Question" is a short story that was written by Isaac Asimov in 1956. If you are interested in reading the story, It is available for purchase in an Asimov collection of short stories called *Isaac Asimov: The Complete Stories, Volume 1.*

At the end of the story, our species has left the picture, and the universe has devolved into chaos. Only AC remains, pondering that one question . . .

I do not want to spoil the ending, but if you read the story, replace the word "Multivac" with "Google."

We are on a path to building superintelligent machines that will think on a much higher level than we could ever imagine. Like Google and Facebook today, these entities will learn directly from us as we engage with them. They will not just interact with us. They will consume us, in a sense, by absorbing all the content we create—all the photos, videos, blogs, emails, posts, tweets, and every other online interaction. That content will contribute to an interactive knowledge base containing humanity's cumulative understanding and wisdom. The process will begin gradually with today's foundation models, but today's pace of innovation is accelerating at an astonishing speed. It seems inevitable that these machines will advance beyond us and, like Multivac, create future versions of themselves.

In the meantime, we need to learn how to coexist and, more importantly, direct, ever-more-capable machines with the ethics determined by society.

In his books and stories, Asimov asked the right questions. If we follow his guidance, we will deeply and permanently encode ethics models within our future intelligent machines that contain the modern equivalent of the laws of robotics and other ethical rules that society determines appropriate.

Asimov showed us the importance of ethics to robots and the intelligent machines we create. I hope we can learn from him. The future of humanity depends on it.

ASIMOV'S LAWS OF ROBOTICS

Zeroth Law: A robot may not harm humanity, or, by inaction, allow humanity to come to harm.

First Law: A robot may not injure a human being or, through inaction, allow a human being to come to harm.

Second Law: A robot must obey the orders given it by human beings except where such orders would conflict with the First Law.

Third Law: A robot must protect its own existence as long as such protection does not conflict with the First or Second Law.

ASIMOV'S LAWS OF HUMANICS

First Law: A human being may not injure another human being, or, through inaction, allow a human being to come to harm.

Second Law: A human being must give orders to a robot that preserve robotic existence, unless such orders cause harm or discomfort to human beings.

Third Law: A human being must not harm a robot, or, through inaction, allow a robot to come to harm, unless such harm is needed to keep a human being from harm or to allow a vital order to be carried out.